The Department of the Environment

Earthquake hazard and risk in the UK

LONDON: HMSO

First published 1993
ISBN 0 11 752773 4

The views expressed in this Report are those of Ove Arup and Partners and are not necessarily those of the Department of the Environment

Contents

Preface

As part of their ongoing programme of Planning Research the Department of the Environment commissioned a preliminary study of UK seismic hazard and risk. This Study was undertaken, principally by Ove Arup and Partners, between 1989 and 1991.

In the UK, as indeed in most parts of the world in regions of low seismicity, the codes of practice do not require any consideration of earthquake loading in the design of conventional structures. For the last two centuries, during which there has been a high standard of historical recording, the quantity of damage caused by earthquakes has been low. The greatest earthquake damage occurred in the 1884 Colchester earthquake where several houses were partially destroyed and there was considerable public alarm. In the same period about 40 earthquakes have caused minor damage, for example to chimneys. Very few casualties have been experienced as a result of these earthquakes.

Until now structural design in the UK has only acknowledged earthquake loading for nuclear plants and other facilities which could give rise to serious problems if failure occured. The issue of earthquake design criteria for conventional structures has arisen in the UK because of the proposal for a standard European seismic code of building practice. Not surprisingly, the prospect of earthquake design rules for conventional structures in the UK, with much lower seismicity than Italy or Greece, has been greeted with concern by UK designers and building regulators since this could have significant cost implications for construction. Nevertheless, the 1989 Newcastle, Australia earthquake showed weaknesses in conventional structures and highlighted the possibility that, even in areas of low seismicity, such as the UK, some basic earthquake design requirements may be necessary.

The work presented herein is a summary of a preliminary study of seismic hazard and risk in the UK. The Study has been completed within a limited time scale and has been based on the best available national information. However, the basic data are of very variable quality and the assessment of seismic risk is, at present, very approximate. The results are specifically aimed at meeting the objectives of the study set out by the Department of the Environment as described below. It is hoped that readers will find this Summary informative but it is important that these generalised results should not be used for purposes other than those for which they are intended. They should, in particular, not be used for seismic hazard assessment of specific sites.

Introduction

This Summary Report has been prepared following the completion of a two year "Preliminary Study into UK Seismic Hazard and Risk".

A Technical Report giving full details of the methods and findings presented here is available from Ove Arup and Partners. Details of the technical report are included on page 32 of this Summary Report

Objectives

The objectives of the Study are:

- to develop an appreciation of seismic risk to the UK insofar as it affects the built environment as a whole;
- to determine whether seismicity should be taken into account in future planning decisions and, if so, how.

Project Team

The Study has been carried out mainly by Ove Arup and Partners in London. Contributions to the work have been made by Delta Pi Associates, London who assisted with seismological aspects and overall methodology and Cambridge Architectural Research who supplied the vulnerability methods for conventional structures. In addition, Earthquake Documentation and Research, Cambridge assisted with verifying the historical earthquake record and Geomatrix, San Francisco advised on methods recently developed for seismic hazard assessment in the USA.

Definitions

The following definitions have been adopted for the purposes of the Study.

Seismic Hazard: The level of ground motion which is expected due to seismic activity.

Seismic Hazard can be expressed in terms of various units, for example, the peak displacement, velocity or acceleration that the ground may experience. Alternatively, it may be expressed in terms of duration of motion or the potential effects of the motion.

Hazard is usually determined for a particular location. It may be the result of a specific earthquake at another location or may be that expected to arise at a certain rate of occurrence due to seismic activity in the region. For example, a designer may wish to know the level of ground motion which has a 10% chance of being exceeded in a 100 year period.

Seismic Vulnerability: The amount of damage experienced by a specific structure arising from a given level of ground motion.

Seismic Vulnerability determines the amount of damage that will be caused by a known level of ground motion.

A structure or building having high vulnerability will suffer significant damage even when subject to a relatively low level of ground motion. Alternatively, a structure with low vulnerability may incur no damage even when subjected to a relatively high level of ground motion.

Vulnerability may be expressed in terms of damage repair cost, or the ratio of damage cost to reconstruction cost. Alternatively, it may be in terms of damage level, for example, the percentage of buildings that are destroyed.

Seismic Risk: The expected amount of damage due to a specific earthquake or which may occur in any given period of time.

Seismic risk may be assessed for individual buildings or for a whole community. It has the same units as vulnerability, for example cost or damage level.

Risk is the combination of hazard and vulnerability and can be used for planning purposes. For example, a measure of risk is the estimated quantity and distribution of damage at a certain locality arising from a specified earthquake. For the purposes of comparing risk it may be useful to express risk in terms of fatalities, for example, "what is the likelihood of 100 people within a given community being killed by an earthquake?".

Hazard * Vulnerability = Risk

Examples of risk

Low Risk:

The structure to the right has very low vulnerability and therefore, even though it is located in a region with high level of hazard the overall risk to this structure is low.

High Risk:

The structure to the far right has suffered significant damage because it's poorly restrained facades are highly vulnerable to moderate levels of earthquake ground motion.

Latin American Tower in Mexico City

School in Newcastle, Australia

Historical perspective of UK earthquake damage

Earthquakes that have caused building damage have occurred many times in the UK. Fortunately there have been very few fatalities as a result of these earthquakes with not more than 8 deaths being recorded since 1580 [21].

Examples of damage reports [25] over the past 900 years include:

Lincoln 1185: "Great stones were rent; houses of stone fell; the metropolitan church of Lincoln was rent from top to bottom."

Wells 1248: "Walls of buildings fell apart and stones were torn from their places in the walls leaving wide gaps and cracks and ruins. The tops of chimneys, battlements and capitals of columns with their architraves were shifted while the bases and foundations did not budge an inch."

English Channel 1580: In Sandwich "it shaked down the gable and coping of the gable end thereof (St. Peters Church) and did shake the cleave tower arches (of St. Mary's Church) and overthrew a piece of chimney. In London, a piece of Temple Church and stones from St. Paul's Church fell down. At Christ's Church stones fell from the top of the church that killed one person and injured others. In the City a little damage occurred to some chimneys."

Inverness 1816: "Chimney tops were thrown down or damaged in every quarter of the town" and "vast quantities of slates and bricks were thrown down." The octagonal spire on the jail "was considerably shattered" and the top 5 or 6 ft. twisted so that "the angles of the octagon are turned nearly to the middle of the .. flat side." The Mason Lodge was "rent from top to bottom ... (the) north stack of the chimney partly thrown down (and) one of the capping stones weighing ... 50 to 60 lbs was thrown ... a distance of not less than 60ft."

A True and Impartial

ACCOUNT,

Of the Strange and Wonderful

EARTHQUAKE,

Which happened in moſt Parts of the City of *London*, particularly, upon the *Royal-Exchange*, *Cornhill*, *Loathbury*, and other Places, between the Hours of One and Two, in the Afternoon, on *Thurſday*, the Eighth of *September*, 1692. To the Terror and Aſtoniſhment of all the Inhabitants, Reſiding, where its wonderful Effects were produced. 9. Sept. 1692.

THAT *Earthquakes* proceed from Air, pent up in the Caverns of the Earth, is the received Opinion of all the Antient Philoſophers, and uncontroverted by all the Moderns, who have made it their Buſineſs to diſcover the internal Secrets of Nature, for the particular Advantage, and Grateful Satisfaction of the moſt Ingenious of Mankind : And becauſe *Earthquakes* are not ſo uſual in *England*, as in other Parts of the World, this was ſo terrible to the Inhabitants, and their Conſternation ſo great and ſurprizing, that all the People were poſſeſt with a Panick Fear ; ſome Swooning, others Agaſt with Wonder and Amaze; the Houſes were deſerted, and the Streets thronged with ſo confuſed Multitudes, that there was no paſſing, nor could any Body give a true Relation of what had paſſed, for a conſiderable time ; ſo great a Terrour and Affright do theſe Subterranean wonderful Cauſes produce, upon the Apprehenſive Minds of Men ; and tho this was not ſo great as has been heard of in Foreign Parts, particularly that which lately happened in *Port Royal*, in *Jamaica*, where ſome Thouſands periſhed ; yet it was ſo ſtrange and wonderful in this Place, that the like has not been known, or heard of, in the Memory of Man ; whether this Prognoſticates any ſtrange future Event, is uncertain, tho ſome has a particular Conjecture in the Affirmative. It being high Change, that Place was deſerted through Fear, and the Merchants and others run, crowding themſelves in the Streets, not daring to venture into the Houſes, ſo great was the Confuſion thereof, to the conſiderable Damage and Detriment of a great many Families, who felt the wonderful Effects of this unuſual and aſtoniſhing trembling ; for in ſome Houſes, in divers Parts, the Pewter and Braſs were thrown from the Shelves, on the Ground ; ſeveral Weavers, as they were at work, in *Spittle-Fields*, and various other Parts, had their work ſpoiled in their Looms ; a Houſe on *Southwark*-ſide, ſunk ſeveral Foot into the Earth, and by the moving of the Earth, many Perſons were taken with a Giddineſs in their Heads, and ſwooning Fits ; divers Chimnies falling down from many Houſes ; this *Earthquake* has not been Circumſcribe within the Bills of Mortality ; but divers Towns in *Surrey*, *Hartfordſhire*, *Eſſex*, *Kent*, &c. hath felt its Effect, particularly the Towns of *Barnet*, St. *Albans*, &c. but leaſt theſe, and other Calamities, are occaſioned by the Sins of this Nation, to avert any Prodigy of this Nature, ought to be the Prayers of all good Men, &c.

London, Printed for J. *Gerrard*, in *Cornhill*, 1692.

South East England 1692

Colchester 1884

Colchester 1884: The damage from this earthquake makes it the most severe earthquake in many centuries. Considerable damage was caused to churches and residential properties. Falling battlements, turrets and parapets were common and the top of one spire also fell. Falling masonry caused substantial damage to church roofs. Reported damage to residential properties repeatedly refers to the falling of chimneys and their penetration through roofs. In brick structures reports repeatedly refer to extensive shattering of walls. In upper storeys falling chimneys caused major damage to ceilings and room contents.

North Sea 1931: This was the largest most damaging earthquake this century affecting areas along the coast of Yorkshire, Lincolnshire and Norfolk. The collapse of chimney pots and stacks was common and the cracking of ceilings frequent. Occasional collapse of ceilings, the dislodgement of roof tiles and cracking of wall plaster occurred. There was one instance of wall and roof collapse.

Carlisle 1979, Lleyn Peninsula 1984, Bishop's Castle 1990 : These earthquakes caused damage to a few chimneys and the occasional more vulnerable structure.

Carlisle 1979

Lleyn Peninsula 1984

The cause of earthquakes

Earthquakes result from a sudden slip of stress across a plane of weakness within the rock mass making up the earth's crust. This process is illustrated below. The two rock masses move relative to each other with time. The rock mass strains and stretches until at some stage the stress across the fault or plane of weakness exceeds the strength of the rock and suddenly the rock slips. This sudden movement releases a large amount of strain energy which radiates away from the source in the form of waves. It is these waves which are felt by observers and which cause damage.

If the two rock masses continue moving relative to each other, the stresses will again build up and the process will repeat itself. The frequency and size of the resulting earthquakes will depend on the rate of relative movement, the stiffness of the rock mass and the strength of the rock across the plane of weakness or fault.

Observation of earthquake locations has shown that while they can occur almost anywhere on the earth, they are concentrated on the tectonic plate boundaries, especially where the plates are in collision. This is illustrated by the two diagrams on the opposite page, comparing the distribution of seismic activity with the locations of the tectonic plate boundaries and their direction of movement . The movements of the plates are caused by convection currents in the semi-molten asthenosphere below the crust. This is illustrated in the schematic W-E section through South America shown below.

On the west coast of South America, where the plates are in collision, and subduction is taking place, many major earthquakes occur.

It is interesting to note in the diagram showing seismic activity that there are several earthquakes, for example in North East China and Central North America, not associated with the plate boundaries. Such events demonstrate that the plates themselves deform and are not devoid of earthquakes. While it is known that earthquakes are concentrated on plate boundaries it is not possible, as yet, to say with any confidence when a major earthquake is likely to occur at any particular location.

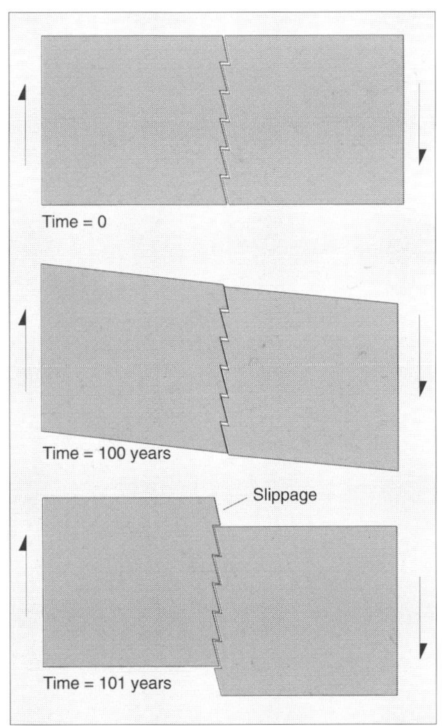

Earthquake mechanism

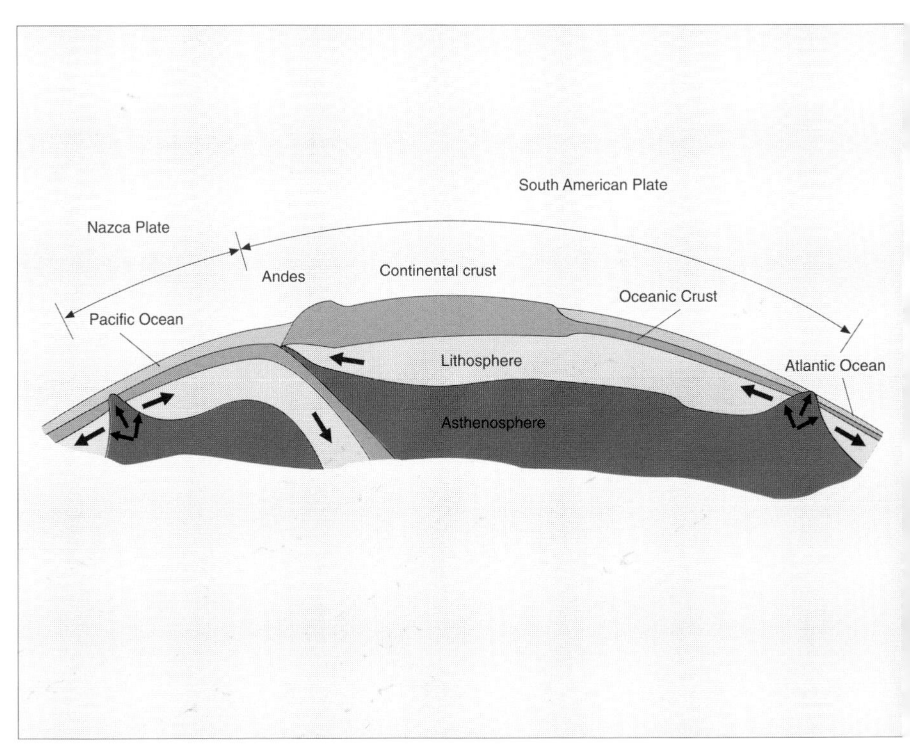

Schematic W-E section through the South American plate

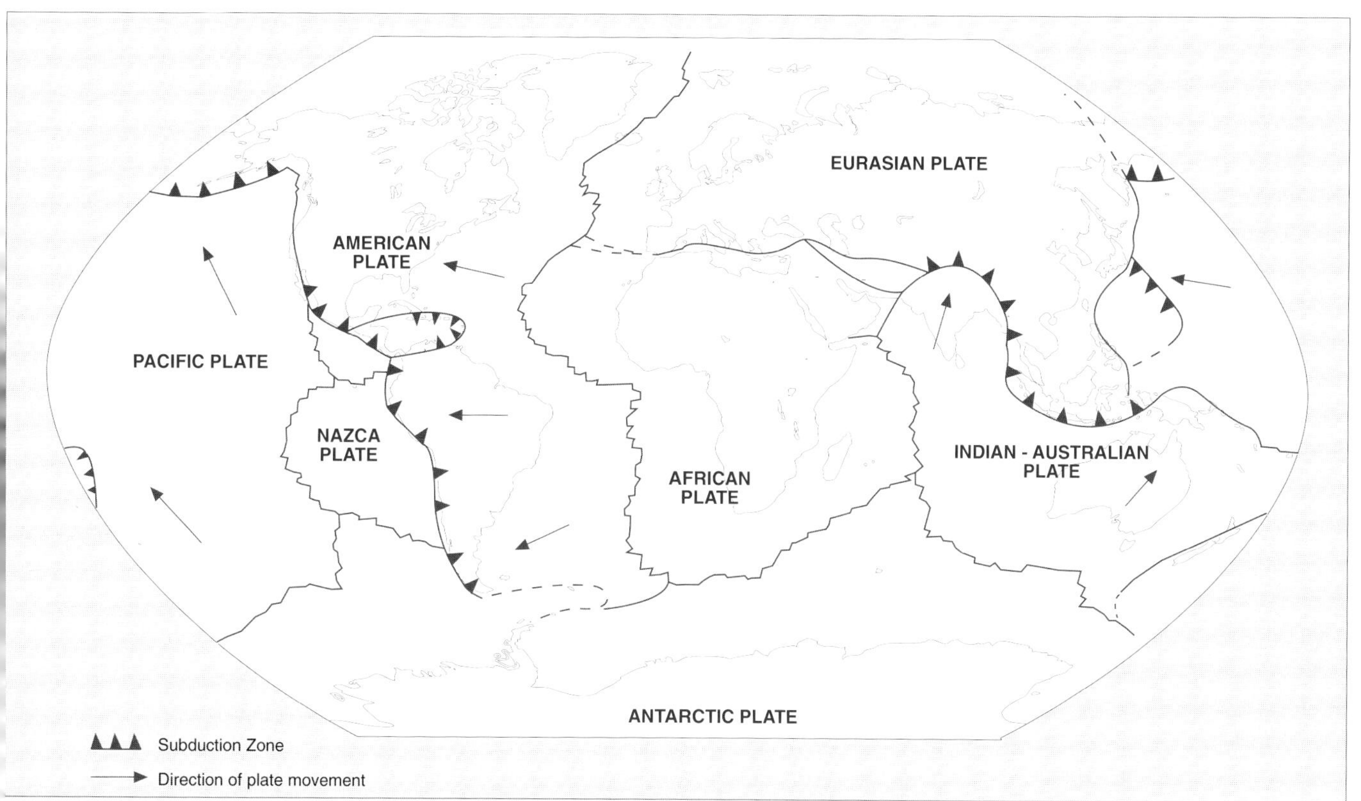

Tectonic plate map of the world

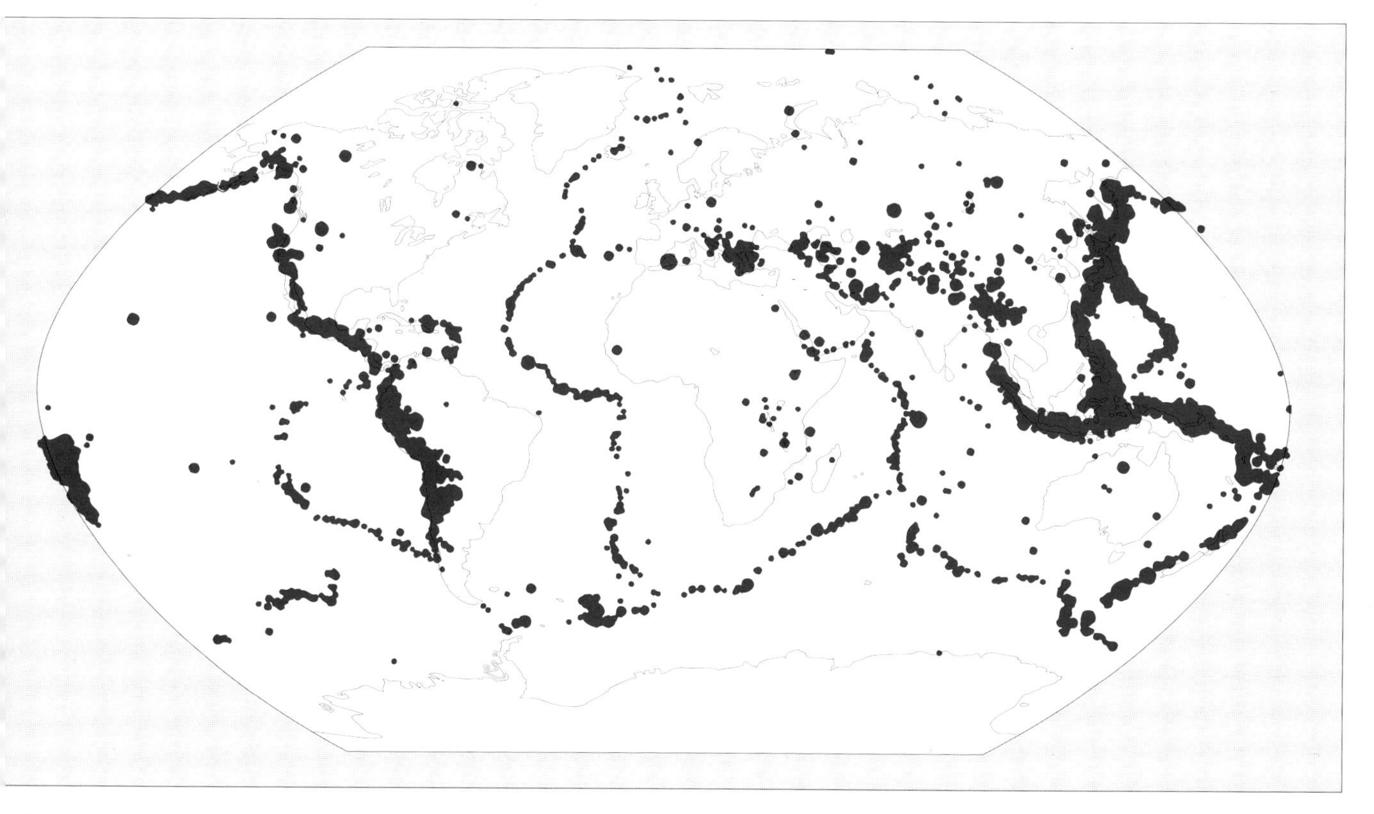

Seismic activity 1980-1989 - Source: British Geological Survey.

The effects of earthquakes

The widely spread effects of an earthquake are due to ground motions resulting from the waves of energy radiating away from the earthquake source. Generally, the damage potential of the waves reduces as they travel further from the source.

Intensity

The effects of an earthquake are expressed using established Intensity Scales (see page 13 for example) with the Intensity reducing with distance as illustrated here

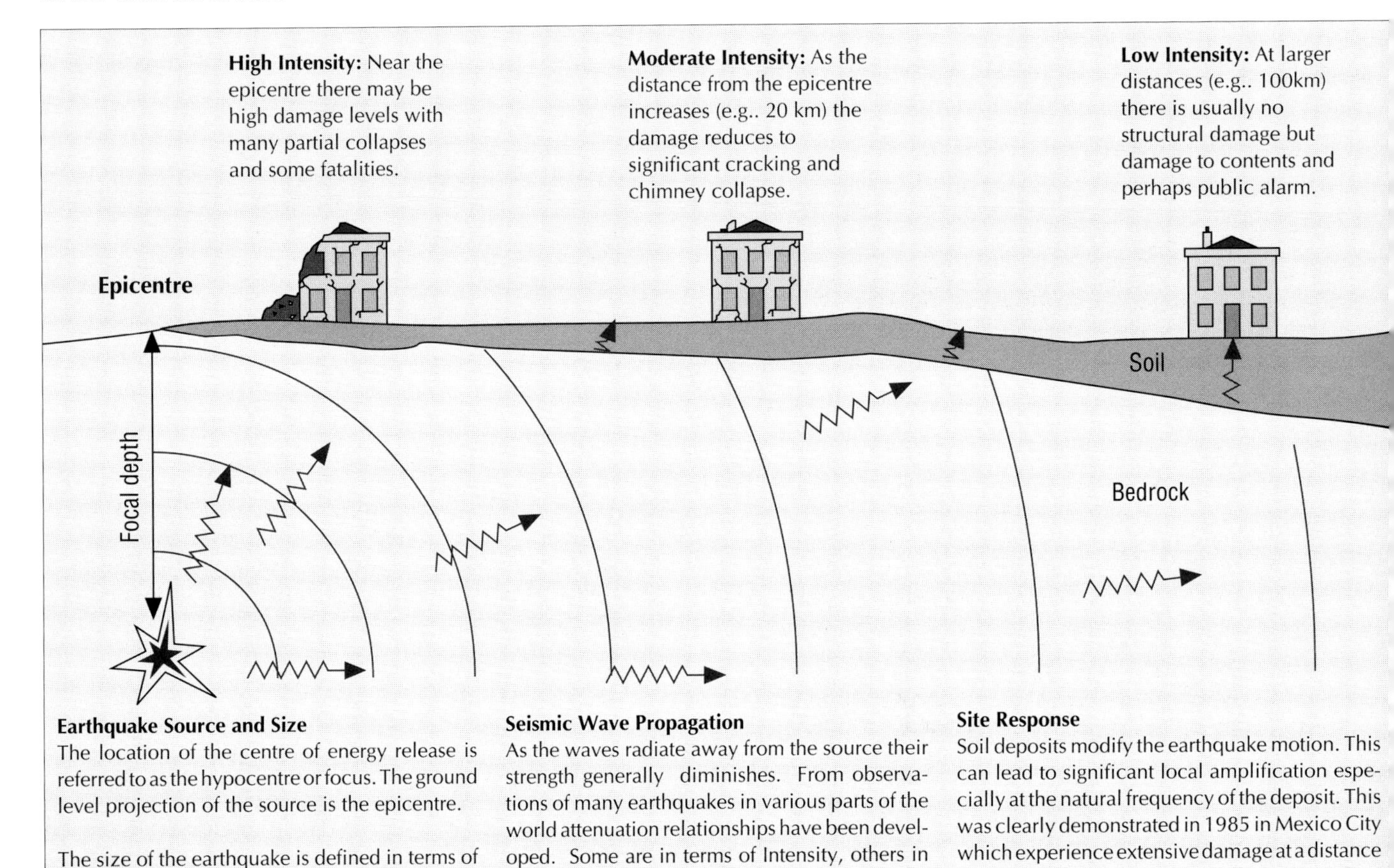

Earthquake Source and Size

The location of the centre of energy release is referred to as the hypocentre or focus. The ground level projection of the source is the epicentre.

The size of the earthquake is defined in terms of its magnitude. Magnitude 4 is quite small whereas magnitude 8 or larger is a great or major event. Each unit increase in magnitude is equivalent to about a 30 times increase of energy released.

Seismic Wave Propagation

As the waves radiate away from the source their strength generally diminishes. From observations of many earthquakes in various parts of the world attenuation relationships have been developed. Some are in terms of Intensity, others in terms of engineering measures such as peak acceleration or velocity of the ground motion.

Site Response

Soil deposits modify the earthquake motion. This can lead to significant local amplification especially at the natural frequency of the deposit. This was clearly demonstrated in 1985 in Mexico City which experience extensive damage at a distance of 450km from the source.

Other Effects

Slope Stability:

In many large earthquakes soil and rock slopes can be caused to fail. Generally, these only occur in conditions where slope stability problems are already known to exist. Earthquake induced slope stability problems are not likely to be significant in the small to moderate magnitude earthquakes expected in the UK.

Liquefaction:

Liquefaction is a phenomenon which can occur in saturated loose sands and silty sands during an earthquake. The sand tends to settle because of the earthquake shaking and is prevented from doing so by the water between the sand particles. This can lead to a quicksand condition and spectacular building foundation failures have been observed in many earthquakes as a result of this. In the UK, liquefaction is very unlikely in natural soil deposits but may need to be considered in loose fill materials.

Consequential Damage:

Fire: Fire can be extremely damaging following earthquakes. In the 1906 San Francisco earthquake, damage to buried gas services combined with mainly timber housing caused a major conflagration. In smaller earthquakes, fires tend to be of less concern.

Leakage: Petrochemical products may leak from damaged vessels/structures with widespread adverse consequences.

Dam Failure: Failure of a large dam in an earthquake could lead to widespread damage and many fatalities.

Methodology used in the study

HAZARD

Seismic activity in the UK: Pages 12 to 15

Seismic activity in a given area is assessed in terms of the annual number of earthquakes greater than a certain magnitude. The available data on instrumental and historical earthquakes as well as the tectonic evolution of NW Europe and the British Isles have been studied and assessed in terms of geographical distribution of seismic activity.

Seismic hazard assessment: Pages 16 to 18

Knowledge of seismic activity and the variation of ground motion with distance from an earthquake source, enables the chance of a given level of ground motion occurring to be calculated. This study has estimated the average seismic hazard level and the variation of hazard around the UK. Since local soil conditions can have a significant effect on the earthquake ground motion their effects have also been considered.

VULNERABILITY

Building stock in the UK. Pages 19 to20

In order to study how vulnerable buildings in the UK are to earthquakes, an appreciation is required of the type and distribution of the building stock.

Vulnerability of buildings: Pages 21 to23

By examining observed damage data from earthquakes around the world for common types of buildings their vulnerability has been established. Vulnerability is expressed in terms of the fraction of the building stock that will experience a certain level of damage for a given level of ground motion. A simplified analysis has been used to check these functions for reinforced concrete buildings.

RISK

Earthquake Impact Studies: Pages 24 to 26

Two areas in the UK were selected and their risk to four specific earthquakes has been assessed. Each area is about 400km^2 and contains a broad cross section of building stock, soil types, land forms etc. Predictions were made for the total cost, the number of structures that would experience various levels of damage and for the number of fatalities. To provide a check of the methodology three earthquakes of known effects have been studied.

Annual costs from earthquakes: Page 27

By combining the calculated seismic hazard with the vulnerability functions annual risk in terms of damage cost as a ratio of reconstruction cost, has been determined for a range of building types. Sensitivity studies have been carried out to determine the relative importance of the various uncertainties in hazard and vulnerability.

Comparison of Earthquake risk to other risks: Page 28

To be able to place seismic risk in context, the predicted annual cost due to earthquakes and number of fatalities were compared with those observed in the UK from other hazards such as weather, fire and subsidence. A very approximate estimate of fatalities has also been made and compared to those observed from other hazards.

Conclusions and Recommendations Page 29

Nuclear installations in theUK are subject to detailed site specific seismic loading considerations and do not form part of this study. Conventional structures however are not designed against seismic loading and the results of this study show that the actual seismic risk is relatively low and therefore this practice is, generally, acceptable. It is inevitable however that there are various existing installations which have higher than average consequences of failure and also structures which are much more vulnerable than average. In these instances seismic design checks may be required.

Seismic activity in the UK

Because of the low seismc activity in the UK it is necessary to consider historical records of earthquakes in addition to recent instrumental data.

Recent Instrumental Data on UK Earthquakes (1980-1990)

Up until the early 1960's, seismological monitoring in the UK was limited to a few non-standard instruments. In the mid 1960's the Institute of Geological Sciences, now the British Geological Survey, installed several instruments near Edinburgh. This network was expanded and by 1979 the instrumental coverage was such that local magnitude (M_L) 2.5 could be detected anywhere within the UK [6]. From the British Geological Survey records the distribution of epicentres of earthquakes, $M_L \geq 2.5$, recorded in 1980 to 1990 is shown here. Two significant earthquakes which occurred during this period were the 1984 Lleyn Peninsula earthquake which had a magnitude 5.4 M_L, and the 1990 Bishop's Castle earthquake, which had a magnitude M_L of 5.1. Both caused some damage to chimneys, etc.

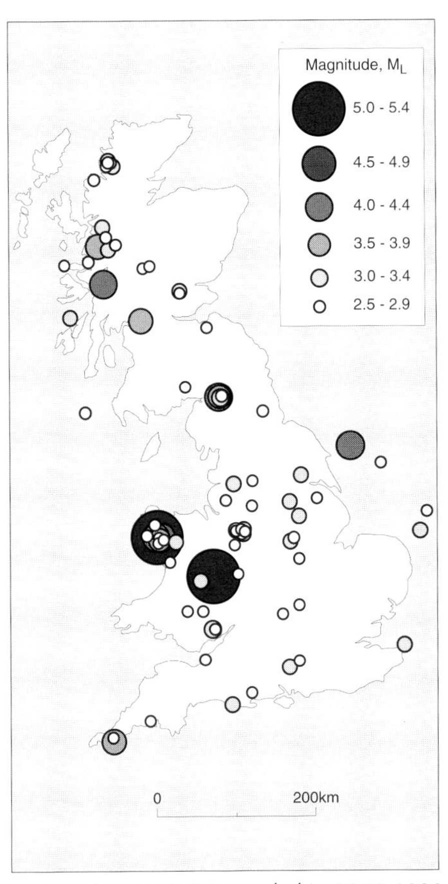

Earthquakes $M_L \geq 2.5$ Recorded in 1980-1990

Magnitude Scales

Instrumental monitoring of seismic activity of an area is achieved using seismographs which measure response to seismic motions. Monitoring is therefore a function of both the disposition of the instruments and their dynamic response characteristics.

The traditional seismological method is based on visual inspection of the various wave forms appearing on a seismogram. The identification of these wave forms enables the epicentre location, focal depth and size of the earthquake to be determined. The size is quantified from the maximum amplitude of a specified wave form on the record after correction for the epicentral distance. The amplitude measured on the record depends on the recording instrument, whereas the correction for the epicentral distance depends on the attenuation from the seismic source.

The original definition of the magnitude (Richter scale of local magnitude M_L) was based on the logarithm of the peak amplitude of a particular instrument (Wood Anderson) in California. The scale is applied to relatively short distances irrespective of the wave form carrying the maximum amplitude and, hence, bears a relationship to the peak ground motion. This scale has been used outside California with and without correction for the local crustal attenuation.

At larger distances the appearance of wave forms depends on the focal depth: shallow events generate strong surface wave signals whereas for deeper events the energy carrier is body waves. After correction for attenuation, measurements of surface waves with a period of 20 seconds (the strongest period in this phase at large distances) give the surface wave magnitude M_S whereas the maximum amplitude of the P-wave phase gives the body-wave magnitude m_B.

The various magnitude scales measure the radiation of seismic waves in certain frequency bands and do not always provide a reliable measure of the size of the earthquake source. A measure of this size is the seismic moment, which is a function of the fault rupture area, the rigidity of the rock and the mean fault displacement. The seismic moment can be estimated in the field and can also be derived from seismograms. Moment magnitude M_W can be determined from the seismic moment. The different magnitude scales are almost equivalent around the value 6 [3].

Historical seismicity

The instrumental data available for the UK covers an extremely short time span when compared with the time scale involved in the recurrence of damaging earthquakes. In order to extend the period of instrumental coverage of UK earthquakes use has been made of the well documented history of the country. By collating "observed" or "felt" effects of earthquakes (macroseismic data), the size and location of a particular event can be estimated.

The "observed" data reported in historical sources is assessed against an Intensity Scale, designed to quantify the effect of earthquake shaking on people, buildings and the natural environment. The Medvedev-Sponheuer-Karnick (MSK) Intensity scale is commonly used in Europe. The MSK Intensity scale is quite explicit in its definitions with respect to buildings and is summarised here.

I	Not noticeable	•	can only be recorded by seismographs
II	Very weak	•	vibration felt by a few people at rest
III	Weak	•	felt indoors by a few people, heavy objects may swing slightly
IV	Largely observed	•	felt indoors by many people, outdoors by a few. A few people awake, windows, doors and dishes rattle
V	Strong	•	felt indoors by most people, outdoors by many. Many people awake
		*	Buildings tremble and pictures sway out of place. Unstable objects may be overturned.
VI	Slight damage	•	felt by most people outdoors. Dishes may break; books fall.
		*	Slight damage to single brick buildings
		*	Chimney damage, small cracks and fall of plaster in a few weak buildings
VII	Building damage	•	Most people frightened, may have difficulty standing
		*	Slight damage to many reinforced concrete buildings
		*	Chimney damage, and small cracks in many brick buildings
		*	large deep cracks in many weak buildings
		•	ground water levels may change
VIII	Destruction of buildings	•	General fright, a few people panic
		*	Large cracks in a few brick buildings and small cracks to many
		*	Partial collapse of many weak buildings and collapse of a few
		•	Steep slopes fail
IX	General damage to buildings	•	General panic
		*	Large deep cracks to many reinforced concrete buildings and a few suffer partial collapse
		*	A few brick buildings collapse, many suffer partial collapse
		*	Many weak buildings collapse
		•	Individual roads and railways are damaged, liquefaction observed
X	General destruction of buildings	*	A few reinforced concrete buildings collapse, many suffer partial collapse
		*	Many brick buildings and most weak buildings collapse
		•	Widespread landslide damage and damage to roads and railways
XI	Catastrophe	*	Most buildings collapse
XII	Landscape changes	*	Practically all structures destroyed
* Indicates building damage			Quantities: a few <10%, many 20-50%, most >60%

MSK Intensity Scale (Simplified)

Assessment of Magnitude of Historic Earthquakes

To enable a seismic hazard assessment to be undertaken the size or magnitude of historical earthquakes must be estimated. Various methods have been devised and usually relate magnitude to either the peak observed intensity or to the area which experiences a certain intensity [1]. Peak intensity is a difficult measure because it is very dependent on the number and type of buildings in the area and on both the magnitude and focal depth of the earthquake. The 1884 Colchester earthquake has produced the greatest damage in the UK in the past 200 years but has been assigned a magnitude of only $M_s = 4.5$ [2]. It has been found that low intensities show less dependence on focal depth than higher intensities. As MSK Intensity IV is the lowest intensity that is consistently likely to be recorded the observed correlation between magnitude and the area affected to at least MSK Intensity IV, for earthquakes where both are known, has been studied. The resulting diagram for UK earthquakes is shown here. The average relationship was used to estimate the magnitudes of earthquakes that have not been recorded by instruments.

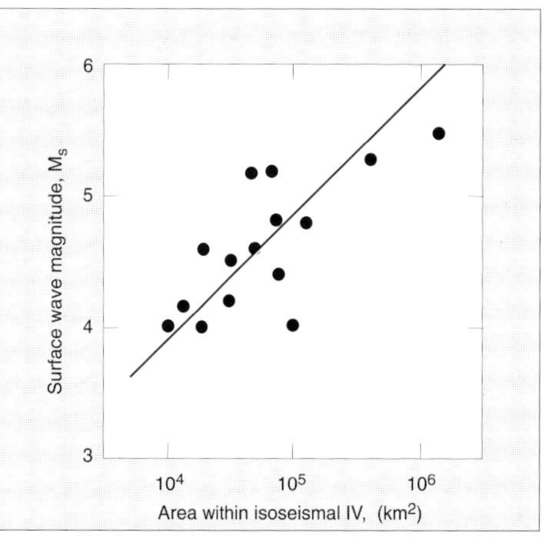

For any particular historical earthquake, the intensity is assessed at the locations mentioned in the region affected, taking negative evidence into consideration. An isoseismal map is constructed by bounding locations of equal intensity within isoseismal lines. The estimated epicentre of the earthquake is generally taken to be either the geographical centre of the mapped area or the location of maximum reported intensity. Isoseismal maps for the 1884 Colchester and 1931 North Sea earthquakes are shown here. Many of these types of maps have been prepared by previous studies carried out principally for the nuclear industry. The studies include those by Principia Mechanica Ltd [23] and Soil Mechanics Ltd [25] in 1982, Ambraseys [2], various studies by Melville [eg. 19], reports by the British Geological Survey [eg. 7] and the site specific studies for nuclear power plants.

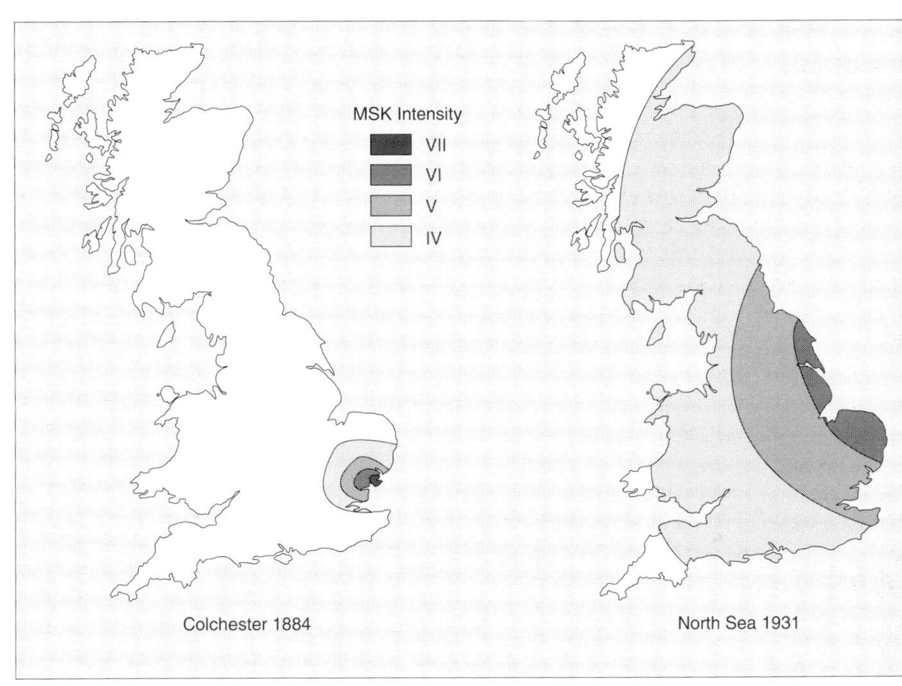

Isoseismal maps [25]

Intensity Recurrence

Seismic activity in the UK can be directly examined by studying the intensity recurrence, or number of times various intensity levels are experienced within any given time period, and how this varies around the country. The diagram to the far right shows the number of times that MSK Intensity VI has been experienced in the UK during the past 200 years. As MSK Intensity VI corresponds to 'slight damage', this diagram shows most areas that have experienced any damage over this time. Many areas have not experienced MSK Intensity VI in this period. This does not mean that they are immune to earthquake damage but that they have not experienced earthquake damage in the past 200 years. The recurrence of MSK Intensity IV gives a better indication of how earthquake activity varies around the country. The diagram to the right shows that MSK Intensity IV has been felt everywhere in the UK in the past 200 years. It can be seen that Wales, the Midlands and parts of Scotland have experienced more seismicity than the remainder of the country.

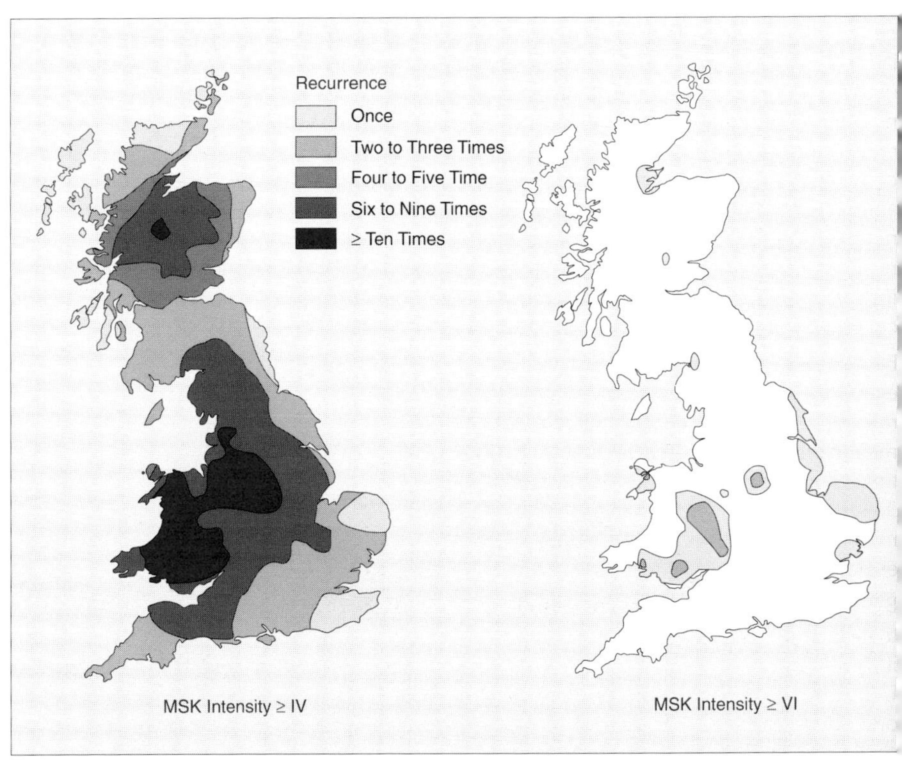

Regional variation of intensity recurrence during the past 200 years

Completeness of the historical earthquake record

The completeness of historical data is influenced by factors such as availability of documentation and population density. Before the advent of modern seismic recording instruments many small earthquakes were not recorded. However, large events would have been felt over a wider area and would be less likely to be missed. Two data subsets of the Historical Earthquake Catalogue are considered to be complete:

M_s≥5, since 1200: A magnitude 5 earthquake should have been detected anywhere in the UK since about 1200. The accuracy of location and magnitude was generally poor before about 1700 however.

M_s≥4, since 1800: Melville [19] drew attention to the importance of regular newspapers and concluded that events with a felt area of approximately 10,000km^2 (about magnitude 4) were certain to have been recorded in England and Wales from 1750, and in Scotland from 1800.

Distribution of historically recorded earthquakes

Earthquake recurrence

To gain an understanding of seismic hazard it is necessary to determine the seismic activity in terms of earthquake recurrence. Generally, there are many more small earthquakes than there are large earthquakes and it is conventional practice to express earthquake recurrence in terms of the annual number of earthquakes greater than a certain magnitude that are within a given area. It is conventional to construct a graph of this annual number against the earthquake magnitude. The diagram to the right shows the resulting graph for the UK derived from the instrumental data on page 12 and the historical data shown above.

Earthquake magnitudes for the historical events are plotted in terms of surface wave magnitude M_S, whereas the recent instrumental data is given in terms of local magnitude M_L. On average there is one magnitude 3.5 earthquake or greater per year and a magnitude 4.5 earthquake or greater about every 10 years.

By plotting similar diagrams for seismic activity experienced in other parts of the world it is possible to gain an understanding of how UK seismic activity compares with that in other parts of the world. The diagram here shows that the UK has an activity rate, for the same area of land, of about a third of the more seismic areas of Eastern USA and less than one hundredth of Japan.

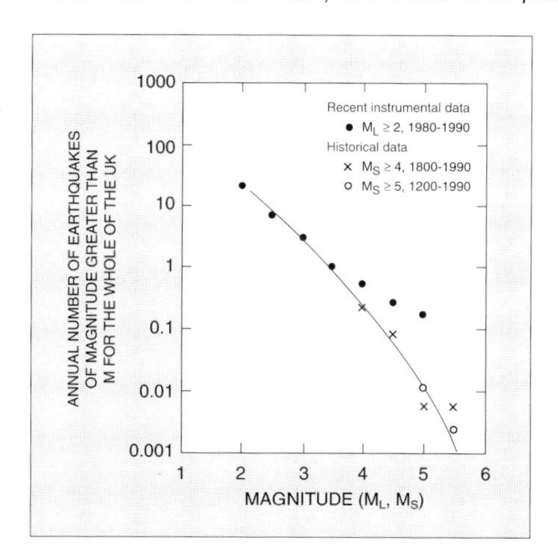

Earthquake recurrence in the UK

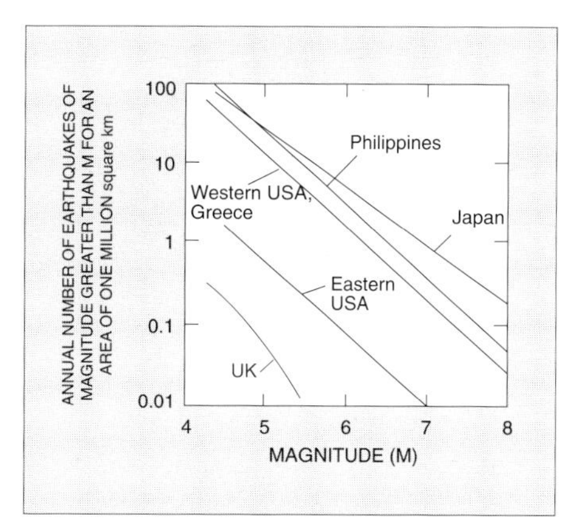

Relative seismic activity rates

Seismic hazard assessment

An assessment of seismic hazard is necessary to determine the rate at which various levels of ground motion will occur.

For the purposes of the Study the ground motion hazard was derived both in terms of MSK Intensity and of horizontal ground acceleration for a site underlain by bedrock. When known fault segments are known to generate earthquakes of similar size a deterministic method can be used for seismic hazard assessment. For regions away from tectonic plate boundaries, such as the UK, this is not the case and a probabilistic method is required. Probabilistic seismic hazard assessments combine the seismic source zoning, earthquake recurrence and the attenuation relationships to produce "hazard curves" in terms of the level of ground motion hazard and the annual rate at which a ground motion of this level or greater will occur [10].

Probabilistic seismic hazard analysis

Information required

To carry out a probabilistic seismic hazard analysis there are two pieces of information that must be assumed.

1.The regional variation of seismic activity. It is commonly assumed that various areas within the total region have a uniform seismic activity. These are referred to as seismic source zones. Each zone is assigned an activity such that the total for all zones equals the total activity of the region.

2. An attenuation relationship. As described on page 10, these are based on observations from previous earthquakes and generally include a measure of variability. Attenuation relationships vary from one part of the world to another but are usually similar in similar tectonic environments. An important feature of the attenuation relationship is the measure of variability associated with it. This variability can have a significant effect on the hazard assessment.

Incorporating uncertainty

Probabilistic seismic hazard assessments have the distinct advantage that uncertainty in the input parameters (source models, attenuation relationships) can be incorporated explicitly [9]. This is achieved by specifying discrete alternatives for parameter values and assigning relative weightings that each discrete alternative is the correct value. For example the source models shown on the opposite page were assigned weightings of 0.3, 0.4 and 0.3 respectively. Uncertainties in focal depth, between 5 and 15km, and earthquake recurrence were also assumed. The earthquake recurrence curves are also assumed to be limited by a range of maximum magnitudes between M_S 5.8 and 6.5.

Methodology

The basic methodology of a probabilistic seismic hazard assessment is as follows. Consider that at any location L it is necessary to calculate the annual rate at which the ground motion will be equal to, or greater than, a particular assumed level of ground motion: H. The following steps are employed.

1. The region is divided into small sub-areas.

2. For each sub-area the number of earthquakes that can occur within any given magnitude range is calculated. This number is equal to the seismic activity per unit area assigned to the seismic source zone appropriate to the sub-area multiplied by the area of the sub-area.

3. The attenuation relationship is used to calculate the likelihood of the level of ground motion H being exceeded as a result of an earthquake of known magnitude occurring in the sub area which is at a distance R from location L. In the example shown to the right the likelihood of H being exceeded is 0.2 for an earthquake of magnitude 4 and 0.7 for an earthquake of magnitude 5.

4. The rate at which ground motion H is experienced is calculated. If the number of earthquakes of magnitude 4 occurring in the sub-area per year is N4 the annual rate of ground motion being at or greater than H is: 0.2 x N4. In a similar way the annual rate from earthquakes of magnitude 5 is: 0.7 x N5.

By adding all annual rates from all of the magnitude ranges the total annual rate arising at location L from the sub-area will be found. By adding the annual rates from all the sub-areas the total annual rate at location L of the ground motion being equal to or greater than H is calculated.

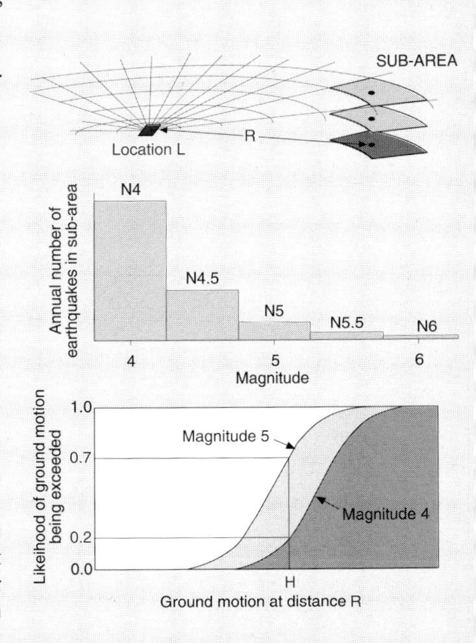

Seismic Source Models

A major uncertainty in seismic hazard analysis within stable continental regions, such as the UK, is the delineation of the boundaries of seismic sources. In an attempt to allow for this uncertainty, three alternative seismic source models have been used as shown below. The total seismic activity was kept the same for all three source models within the whole area A. The 3 models are:

Model 1: Average Seismicity - A uniform seismicity is assumed over the whole area A.

Model 2: Zoned Seismicity - This model includes three zones of higher activity. Similar seismic activity in terms of number of earthquakes per unit area was assumed in these zones and a lower background activity was assigned to the surrounding area. It is primarily based on the locations of earthquakes, $M_S \geq 4$, that have occurred since 1800.

Model 3: Tectonic - This source model was based on deformation phases over the past 50 million years [20] and the activity of the tectonic zones was defined from the observed rate of the earthquakes with magnitude $M_S \geq 5$.

Attenuation Relationships

Attenuation relationships have been chosen for two hazard parameters:

MSK Intensity the MSK Intensity attenuation relationship developed for north west Europe [1] has been used. This macroseismic attenuation is based on isoseismals that envelope, rather than average observations. To obtain a "mean" relationship, the attenuation has been lowered by one half of an intensity unit. Variability in the attenuation relationship has not been published but for the Study a standard deviation of MSK Intensity of 1.0 was assumed. This intensity attenuation relationship has been derived from data covering all types of soil conditions. It therefore incorporates the 'average' effect of local soil amplifications.

Peak horizontal ground acceleration the attenuation relationship recently derived for regions remote from tectonic plate boundaries [11] has been used. This relationship is for bedrock sites.

Minimum Magnitude

A minimum magnitude, M_S=4, has been used in the hazard calculations because the historical record shows that the likelihood of an earthquake with a smaller magnitude causing any more than trivial damage is extremely low.

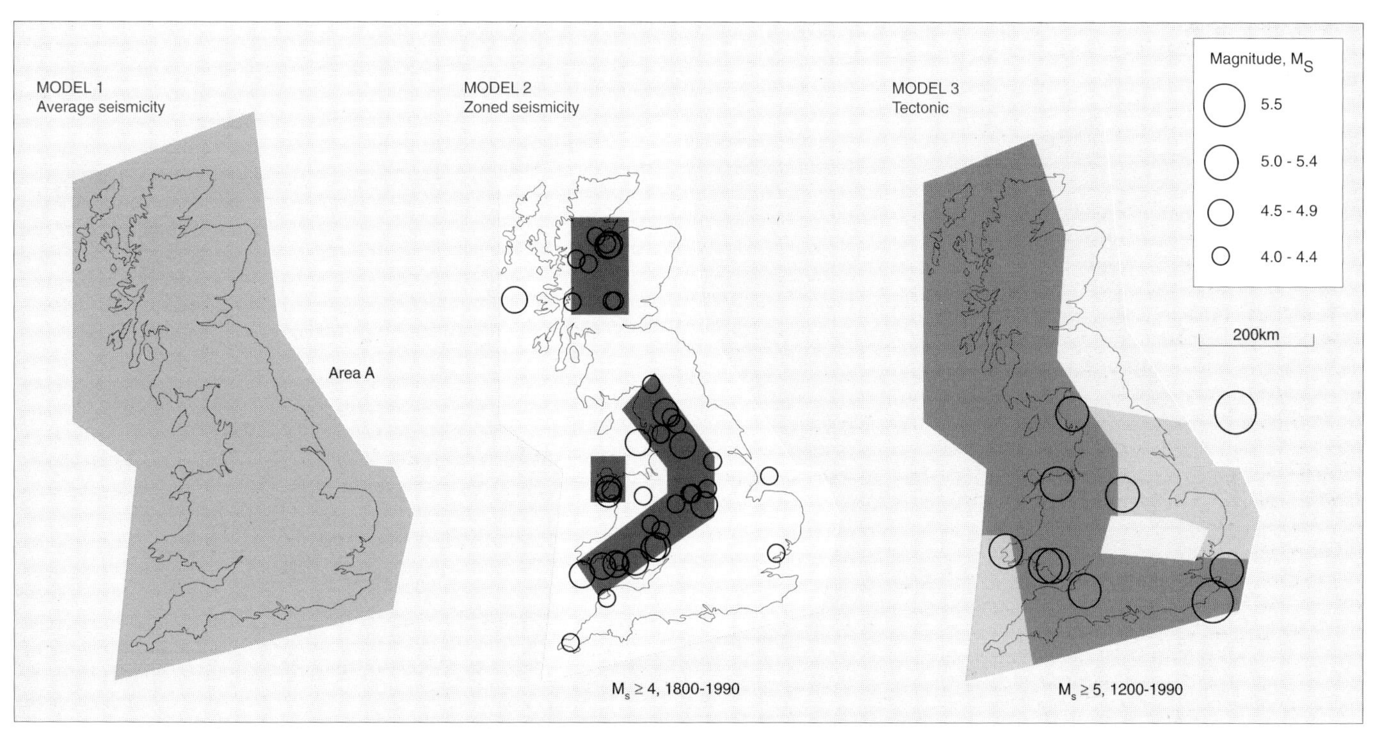

Seismic Source Models used in Seismic Hazard assessment

Calculated hazard levels

The level of seismic hazard has been calculated for two parameters

MSK Intensity

The average calculated intensity recurrence curve for the UK is shown to the right. It shows that, on average, at any particular location in the UK, the calculated annual rate of occurrence of MSK Intensity VI is 1 in 1,000 and for MSK Intensity VII is 1 in 10,000. The calculated regional variation of MSK Intensity recurrence is shown on the far right.

Peak horizontal acceleration

The response of structures to horizontal acceleration in an earthquake is a function of the fundamental period of the structure. Low rise structures tend to amplify the motion whereas high rise and other long period structures experience reduced acceleration. The diagram to the bottom right shows the avarage calculated peak horizontal acceleration against structural type for bedrock ground motion having an annual rate of occurrence of 1 in 500.

Site response effects

It is well known that local soil effects can have a significant effect on earthquake damage levels. The damage in Mexico City in 1985 was an extreme example and soil effects were very evident in the 1989 earthquake which damaged San Francisco. The MSK Intensity hazard calculations incorporate the average effect of local site amplification. Horizontal acceleration, however, can be modified explicitly to allow for effects of local soil amplification.

Researchers at McMaster University, in Canada, have recently carried out a systematic study of site response effects using the methods described in [15]. Their results show that amplification for soft clay is greater than other soil types especially at fundamental periods greater than 0.4 seconds. At lower periods, however, the amplification is greatest for a small thickness of stiffer soil. This implies that conventional one to three storey structures will experience more earthquake damage when underlain by relatively thin soil deposits. This was observed in the 1971 San Fernando earthquake [24] and in the 1989 Newcastle, Australia earthquake [12].

The calculated effect of a 5m thick stiff clay deposit, for example, is shown here.

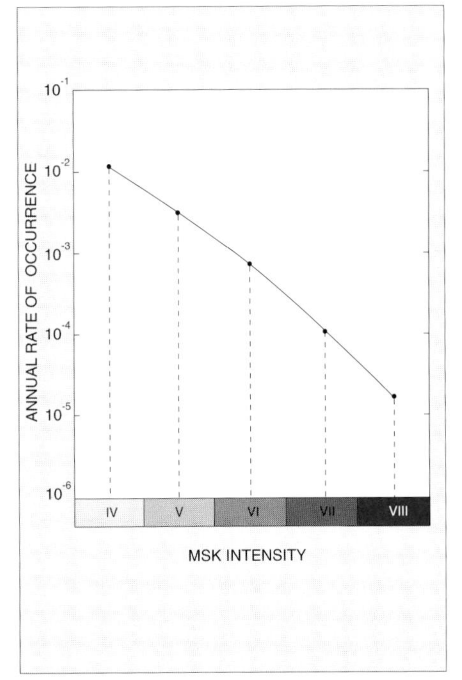

Calculated average rate of MSK intensity for the UK.

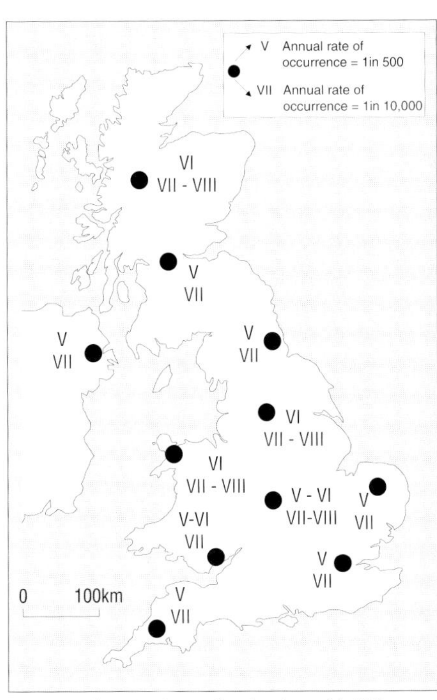

Regional variation of calculated MSK Intensity recurrence

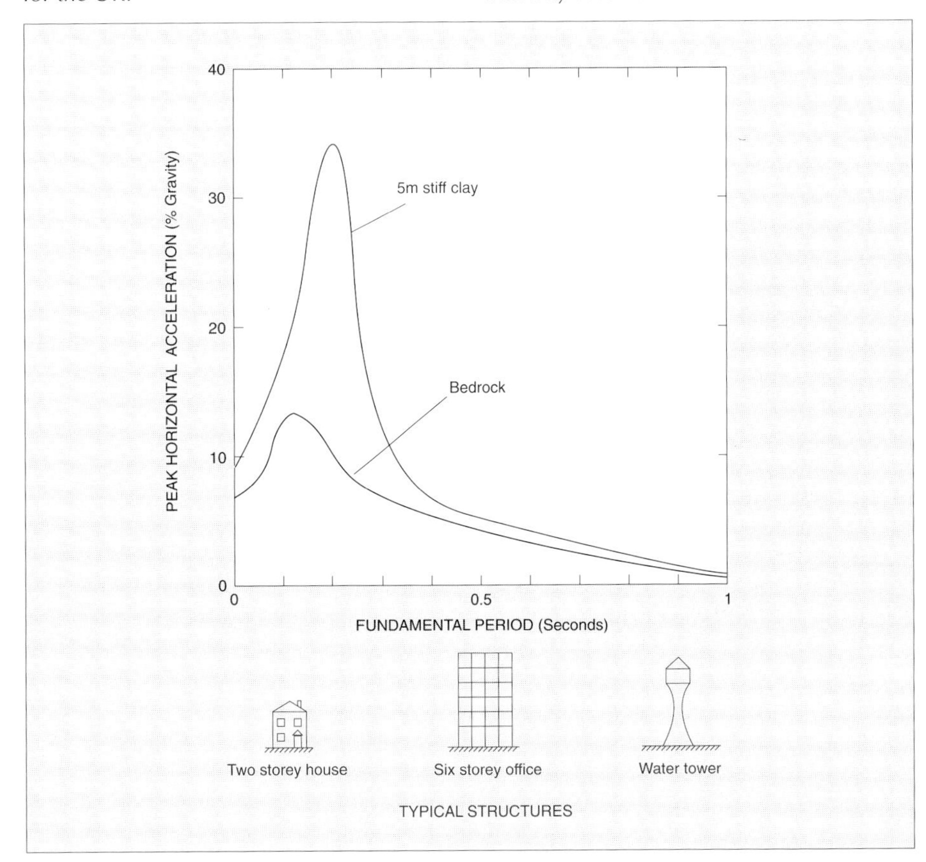

Typical horizontal acceleration having an annual rate of occurrence of 1 in 500

Building stock in the UK

In addition to considering the likelihood of earthquakes occurring in the United Kingdom, it is important to review the building stock that would be affected by them.

The United Kingdom is one of the most densely populated parts of Europe and the numbers of buildings and concentrations of infrastructure, industry and towns means that an earthquake occurring almost anywhere in Britain will affect populated areas. Four of Britain's cities (London, Birmingham, Manchester and Leeds) are among the 100 largest cities in the world.

Much of the urban development and industry in the United Kingdom has been built over the past century. Earthquakes that affected the UK in the past, like the Colchester earthquake of 1884, described on page 7, occurred when the population was less than half of its current level and when the value of the building stock and infrastructure was only about 10% of present day values. A similar sized earthquake occurring almost anywhere in Britain today could affect many more people and buildings.

To assess the likely effects of earthquakes, the building stock has been broken down into categories of building type. Building types were classified by their use, construction type, age and size. The construction type of the building stock of the United Kingdom is very uniform: 90% of houses are brick masonry with slate or tiled roof. A similar percentage of smaller non-residential buildings are also of this traditional construction. Larger buildings (those greater than 300 m^2) are mainly engineered structures, built from reinforced concrete frame or, less commonly, steel frame, precast concrete panel or other structural systems.

The breakdown of the national building stock is given in the table to the right. Although this national picture is typical of the building stock in many parts of the country, the distribution can vary from place to place. The age profile of the building stock varies locally, but regionally is fairly uniform across Britain.

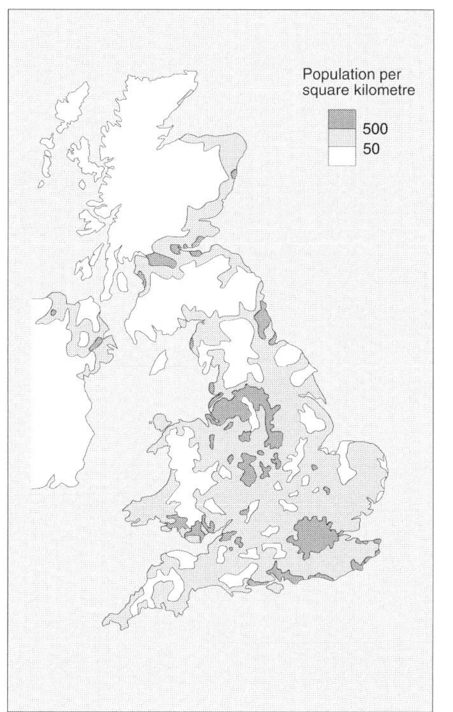

Population distribution

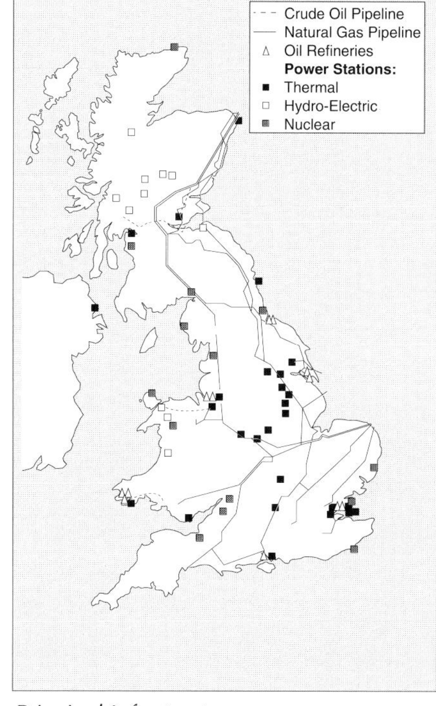

Principal Infrastructure

Residential	Total floor area in UK (million square metres)	As a proportion of the total	Floor area per person (square metres)
Houses, Masonry, before 1920	439	13.6%	7.4
Houses, Masonry, 1920 - 1940	923	28.6%	15.6
Houses, Masonry, after 1940	692	21.4%	11.7
Other Construction Types	254	7.9%	4.3
Total Residential	**2,308**	**71.4%**	**39.0**
Non-Residential			
Offices	148	4.6%	2.5
Industrial	273	8.4%	4.6
Health	14	0.4%	0.2
Retail	107	3.3%	1.8
Hotels	27	0.8%	0.5
Catering and Entertainment	96	3.0%	1.6
Churches	21	0.6%	0.4
Warehouse	132	4.1%	2.2
Education	105	3.2%	1.8
Total Non- Residential	**923**	**28.6%**	**15.6**
Total Building Stock	**3,231**	**100%**	**54.6**

Breakdown of Building Stock in the UK

Likely earthquake damage characteristics

The types of buildings that exist in Britain today are very different to those that made up the 19th century cities. Many more of today's buildings are professionally engineered and 20th century building regulations and design codes have resulted in continually improving construction standards. Today's buildings are likely to be more robust and generally less vulnerable to earthquake effects, even though very few have been specifically designed to resist earthquakes.

Data from past earthquake damage in the UK and from earthquakes affecting similar, buildings in other countries have been used to predict damage levels in the current UK building stock. It is clear from surveys of damage in recent earthquakes in Britain and else where that the older masonry buildings, and those in the poorest condition, are the most vulnerable.

Brick masonry buildings built before 1920 constitute nearly 20% of the residential buildings of this country. Those that are rundown, in a poor state of repair and already deteriorated are the most likely to suffer damage in future earthquakes. Buildings already cracked or suffering from subsidence are likely to have existing damage made worse. Domestic chimneys are the most common locations for earthquake damage, particularly poorly maintained chimneys, tall stacks and those weakened by weathering. Parapets and gable walls are also likely to be places where damage occurs.

Typical 19th century industrial building

Recent multistorey residential building

Typical pre 1920 housing

Calculating costs

Earthquake damage costs depend on the types of buildings affected, their vulnerability, size and the costs of repair and reconstruction. Non-residential buildings tend to be large structures and more costly to build per square metre than residential buildings. Typical construction costs for houses are around £600 per square metre but for offices, shops, hospitals and other non-residential facilities are over £1,000 per square metre. Industrial facilites and warehouses are less expensive to build: around £500 per square metre. These factors have to be taken into account in assessing the structural costs of earthquake damage.

Typical recent housing

Vulnerability of buildings

Vulnerability functions for the most common UK building types have been defined using the data and methodology of Cambridge Architectural Research [26]. These are based on data from a large number of earthquake damage surveys around the world and have been compared with other published data and basic structural calculations.

The use of damage surveys

A group of buildings of the same type (for example brick masonry built 1920-1945) in a district shaken by an earthquake will experience a range of levels of damage among the structures. This distribution of damage, for example what percentage of buildings suffer slight damage when others suffer moderate damage, is fairly predictable and depends on the severity of shaking, or intensity at that place. Earthquake intensity scales (for example the MSK Intensity Scale on page 13) are partly defined from the distribution of damage to certain common building types.

The collection of a large number of damage surveys enables expected distributions of damage to be defined for varous levels of earthquake shaking. It also enables the relative vulnerability between different building types to be derived.

The Newcastle, Australia earthquake

An example of an earthquake damage survey to buildings similar to those in the United Kingdom is the data from the 1989 Newcastle, Australia earthquake [12]. Here a survey of the damage to 620 buildings was carried out from photographs of every building within a given defined area. The damage levels, D1 to D5, were assessed in accordance with the table above. The diagrams to the right show the observed damage as a percentage of the total building stock. It can be seen that there is a noticeable influence of age on the damage levels to masonry buildings.

The damage in Newcastle cannot be directly related to ground motion as there were no ground motion instruments in the vicinity. The only data available are the percentage of buildings damaged to each damage level and their relationship to each other.

Damage to masonry building in Newcastle, Australia 1989

Damage Level	Definition for Loadbearing Masonry	Definition for Reinforced Concrete Buildings	Repair Cost Ratio*
D0 Undamaged	No visible damage	No visible damage	0
D1 Slight	Hairline cracks	Infill panels damaged	0.05
D2 Moderate	Cracks 5-20mm	Cracks 10mm in structure	0.20
D3 Heavy	Cracks 20mm or wall material dislodged	Heavy damage to structural members, loss of concrete	0.50
D4 Partial Destruction	Complete collapse of individual wall or individual roof support	Complete collapse of individual structural member or major deflection to structure	0.80
D5 Collapse	More than one wall collapsed or more than half of roof	Failure of structural members to allow fall of roof or slab	1.0

*Repair Cost Ratio: Ratio of the cost of repair of the damage to the cost of reconstruction.

Damage level definitions

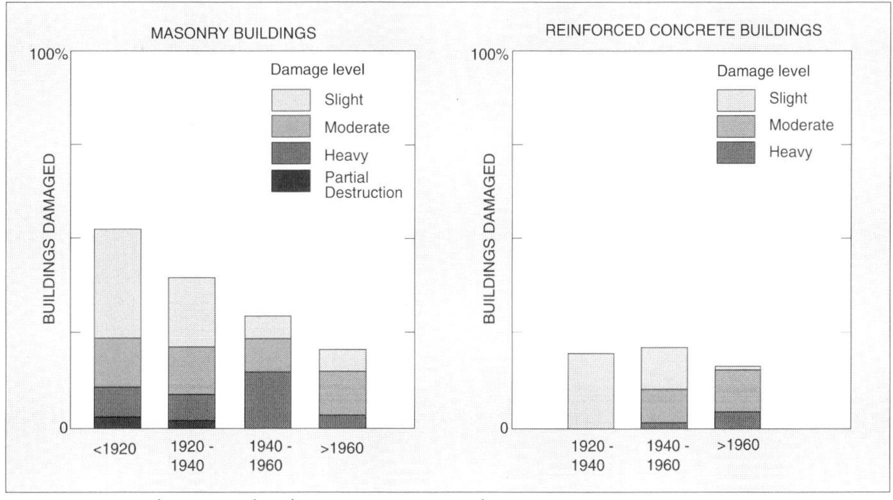

EEFIT Newcastle, Australia damage survey results [12]

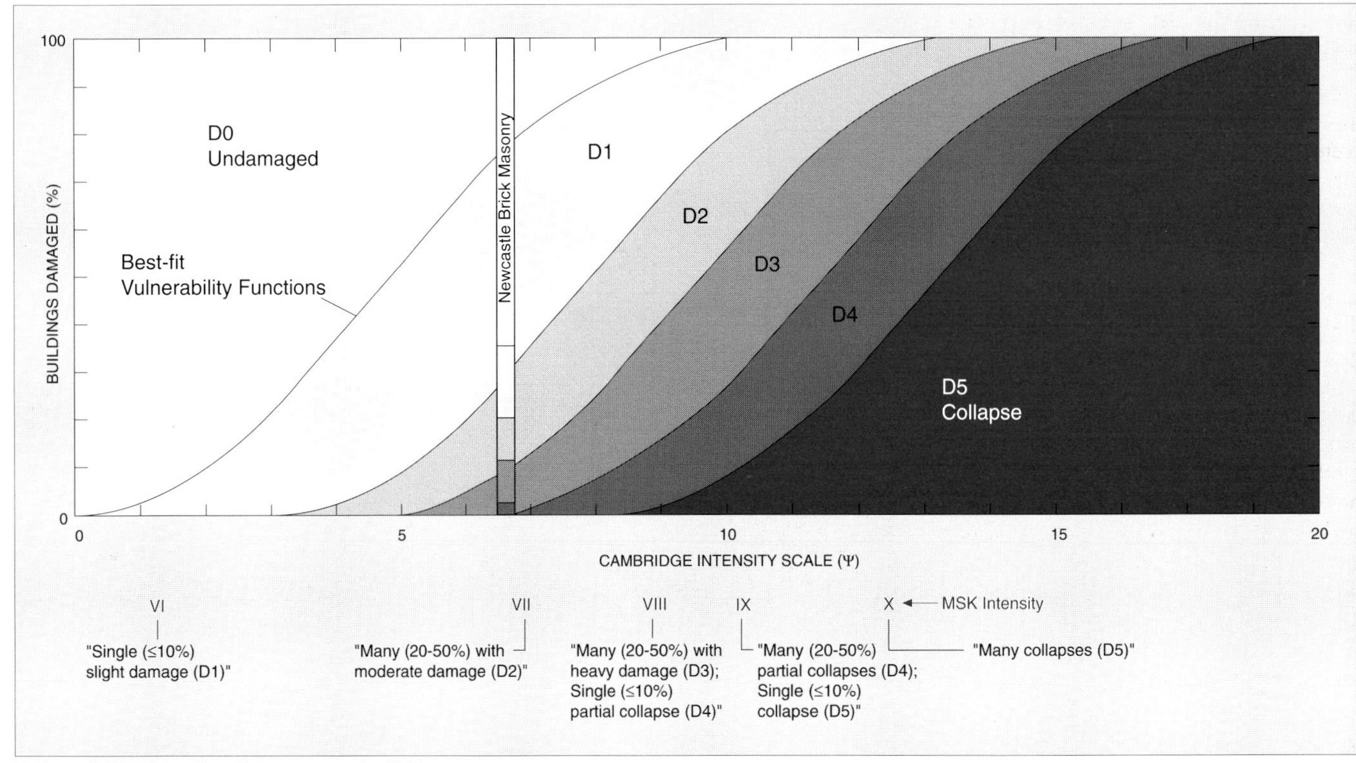

Vulnerability function for masonry buildings

Brick Masonry Buildings

Examination of a large data set (103 surveys of damage to brick masonry buildings) shows that there is a consistency between the various damage levels for all the surveys. In other words, if the percent of buildings damaged to at least any one damage level is known then an estimate of the percent of buildings to experience at least any other damage level can be made with some confidence.

The diagram above shows a set of continuous curves defining the percent of buildings experiencing the various damage levels. The horizontal scale is referred to as Cambridge Intensity Scale ψ. The Newcastle masonry data is highlighted to show what comprises a data entry. Each data entry consists of the percentage of the buildings damaged to each damage level (D1 to D5). The horizontal position of each data entry has been moved, to obtain the "best fit" of all five damage levels with those of other surveys. The continuous curves are based on the "normal distribution" function (the most likely function to be appropriate) and fit the data relatively well.

Relating ψ to MSK Intensity

Hazard levels for the UK have been determined in terms of MSK Intensity as described on page18. To calculate risk using the ψ scale means a relationship between ψ and MSK Intensity must be derived. The definitions of MSK Intensity between VI and X are relatively precise for damage levels to 'Ordinary Brick Houses'. These definitions have been used to superimpose the MSK Intensities on the ψ scale above at the value where the level of damage best agrees with that indicated by the scale.

These results have been compared to those produced by the US regulatory research body, the Applied Technology Council [4]. They have carried out an "expert opinion" survey of damage levels that may be expected for a given intensity level. For low rise brick masonry for example they have produced a best estimate and high and low estimates of damage. For Intensity VII or greater they agree quite well with those derived using the ψ methodology. For Intensity VI, the ψ methodology predicts significantly lower levels of damage.

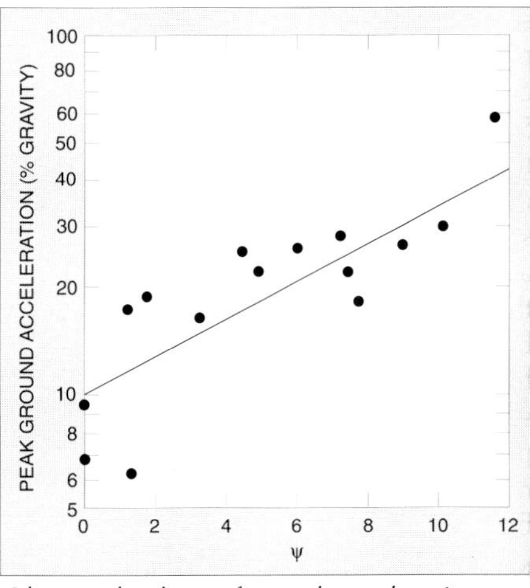

Observed values of ψ and acceleration

Relating ψ to Ground Motion

Several past studies have obtained very approximate relationships between ground motion (mostly peak horizontal acceleration) and various intensity scales. As part of the Cambridge damage surveying methodology over the past ten years 14 damage surveys have been carried out close to ground motion recording instruments [26]. The correlation of ψ with ground motion parameters has been studied. Acceleration parameters generally correlate with ψ better than velocity or duration. The correlation between peak horizontal ground acceleration and ψ is shown above.

Reinforced Concrete Buildings

Reinforced concrete buildings have been considered in a similar way to masonry using the ψ Intensity scale. The reinforced concrete curves are similar to those of masonry but are about 2 ψ units more earthquake resistant. The Applied Technology Council [4] assessments for low rise reinforced concrete non-ductile frame structures show a similar damage to intensity relationship to that indicated by ψ. By using the ψ to peak ground acceleration relationship shown in page 22 and the repair cost ratios in the table on page 21 a curve for damage cost as a proportion of reconstruction cost can be plotted against ground acceleration as shown here. Hwang and Jaw [16] calculated the seismic vulnerability of a 5 storey reinforced concrete shear wall structure and have expressed their results in terms of damage level against peak ground acceleration. Their results are seen to predict higher damage at low acceleration levels.

In order to check the vulnerability of reinforced concrete buildings simplified calculations have been carried out which relate seismic vulnerability to the minimum lateral strength required in UK codes of building practice. Data on vulnerability given for code designed buildings given by the United States National Earthquake Hazard Reduction Program (NEHRP) [22] have also been considered. A lower bound ultimate lateral strength of UK buildings over 2 storeys was assumed to be 1.5% of deadweight which is the current UK code requirement. An upper bound lateral strength has been assumed to be twice the lower bound strength. Recent studies [5] suggest that wind loads would probably impose higher loads than the minimum requirement, for reinforced concrete buildings over about 20 storeys.

NEHRP [22] specifies the ultimate lateral strength required to withstand a given level of design earthquake acceleration. It also gives an estimate of the level of damage as a function of the peak acceleration experienced in an earthquake as a ratio to the design acceleration. Using these relationships, the damage resulting from a given acceleration can be estimated for both the lower bound and upper bound strengths. The results shown here are seen to bracket those derived from ψ and those calculated by Hwang and Jaw. It is considered that this exercise lends credence to the ψ vulnerability curves. However some structures, particularly those with significant irregularity, may exist that correspond to the lower bound strength and consequently have much higher than average vulnerability.

Collapsed reinforced concrete building, Mexico City 1985

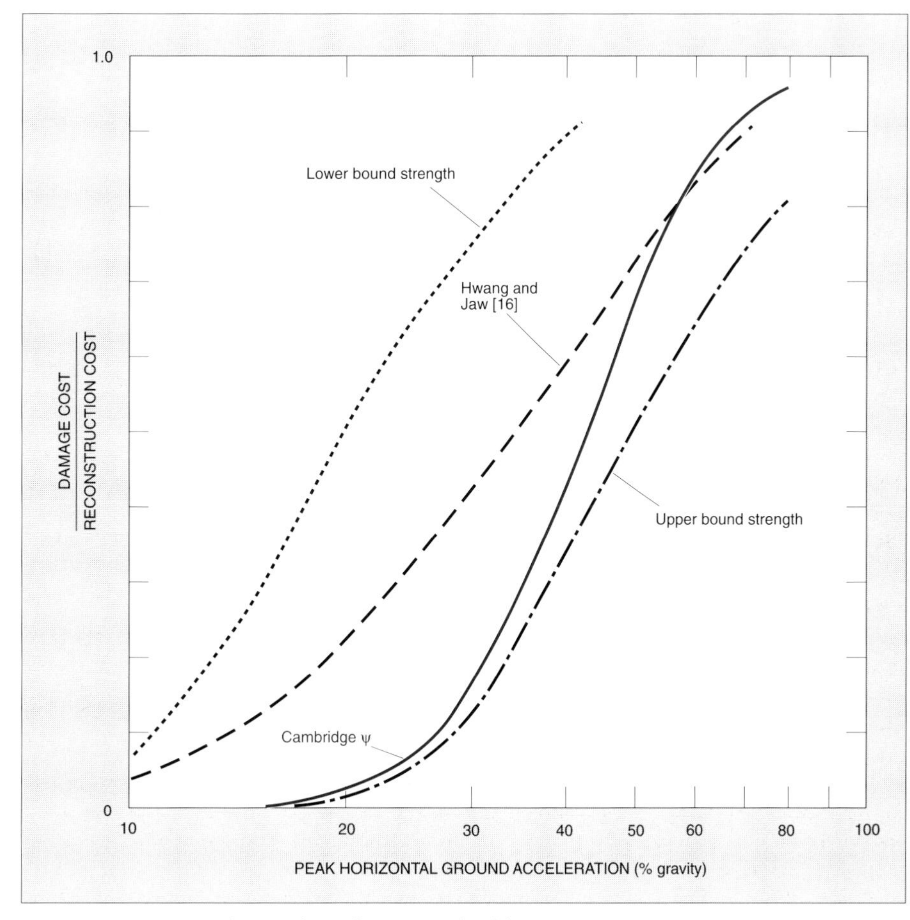

Damage cost ratios for reinforced concrete buildings

Earthquake impact studies

For the purposes of the Study two representative areas in the UK were chosen. They had above average seismic hazard, an area of about 400km^2 and a mixture of urban and rural land use. For each area (A and B) calculations have been carried out to estimate the effects of four specific earthquakes.

This calculation combines the attenuation relationship, the soil amplification (if appropriate), the distribution of building types and their vulnerability functions. Two sets of calculations were performed, one using MSK Intensity as the hazard parameter and the other using peak horizontal acceleration corresponding to a structure with a fundamental period of 0.2 seconds.

Specific earthquakes

To provide a check of the methodology three earthquakes of known effects have been studied. They are:

The **1990 Bishop's Castle, UK earthquake** A magnitude 4.4M_S with a focal depth of 14km.

The **1983 Liege, Belgium earthquake** A magnitude 4.5M_S with a focal depth of 4km.

The **1989 Newcastle, Australia earthquake** A magnitude 5.5M_S with a focal depth of 10km.

An **extreme earthquake**, (by UK standards), a magnitude 6.5M_S earthquake with a focal depth of 15km was also modelled.

Geographical information

Computer maps of both areas were assembled and the geographical data shown in the diagram were incorporated as follows:

- Soil Type
- Ward and Parish boundaries
- Current Land Use
- Historical Development

Using population statistics from census data the number and distribution of the residential building stock could be estimated from the Ward, the land use and the historical development period. The number and type of non-residential buildings were estimated in a similar way but many assumptions were required as the available data is generally only in terms of national totals (see page 19).

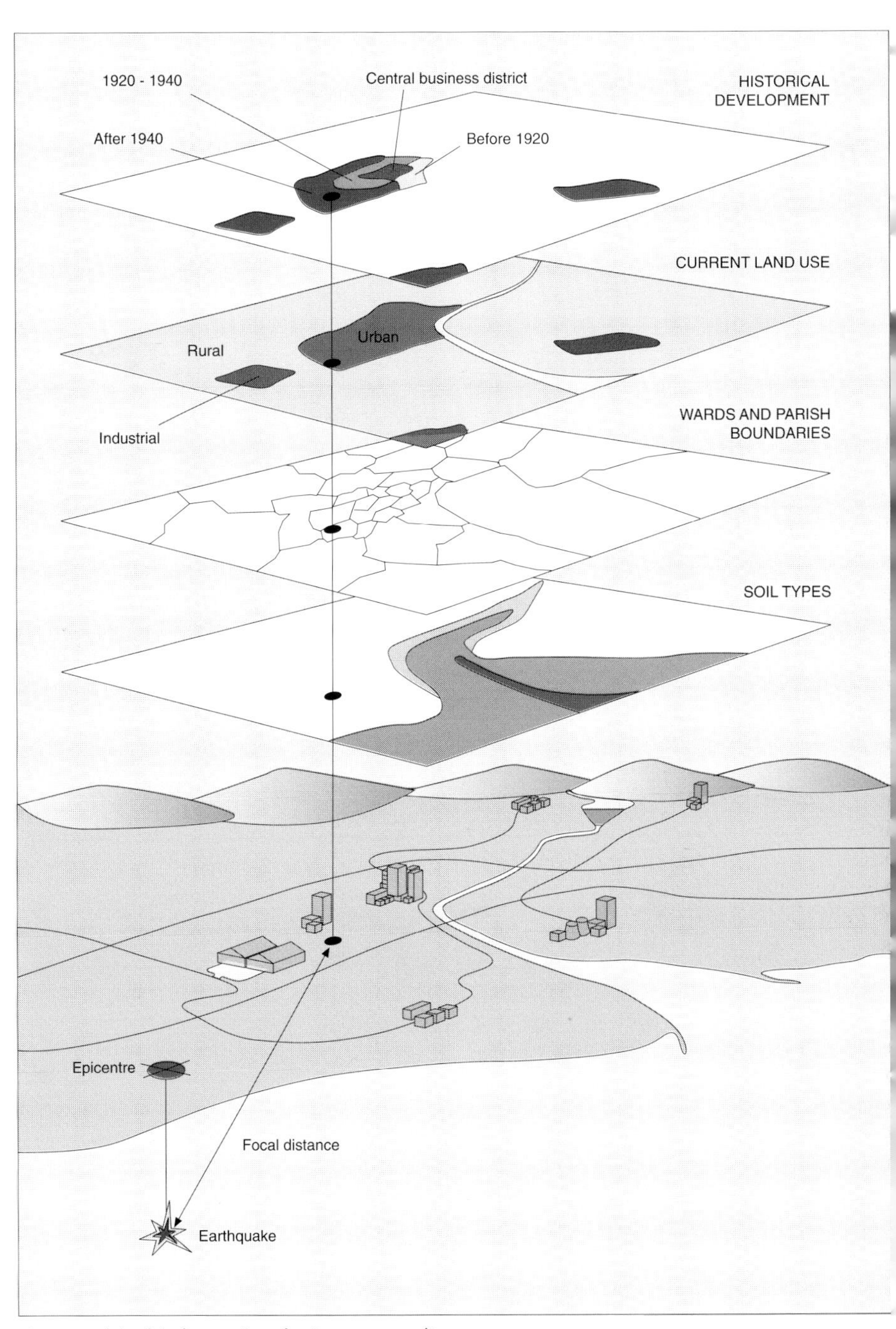

Geographical information for impact studies

Predicted building damage

The predicted quantity of damage for the 5 levels of damage D1 to D5 as a result of each of the four earthquakes are shown here in terms of percentage of the total building stock. The results are as follows:-

Bishops' Castle type earthquake

The calculation using MSK Intensity as the hazard parameter gives no damage whereas the acceleration calculation shows about £10 million. The annual rate of occurrence of an earthquake of this size within each area is about 3 in 10,000.

Liege, Belgium type earthquake

The calculated cost is between £150 and £250 million. The actual cost of damage due to the Liege earthquake was about £50 million [17]. These costs can be compared as Liege has a similar population to the microzoning areas. The annual rate of occurrence of a shallow event of this magnitude is about 1 in 10,000.

Newcastle, Australia type earthquake

The calculated costs vary between about £1.2 and £1.7 billion. The actual reconstruction works of the Newcastle earthquake amounted to about £1 billion. Again Newcastle has a similar population to the study areas. The majority of housing in Newcastle comprised single storey bungalows of which about 25% are brick masonry and 75% are timber. If the housing had been 2 storey brick masonry as in the UK, losses would have been higher. The annual rate of occurrence is about 2 in 100,000.

Extreme earthquake

The calculations show extensive damage with total costs of about £4 billion which amounts to about half the total value of the building stock. The annual rate of occurrence of such an event occurring in the vicinity of each area is less than 1 in 10 million.

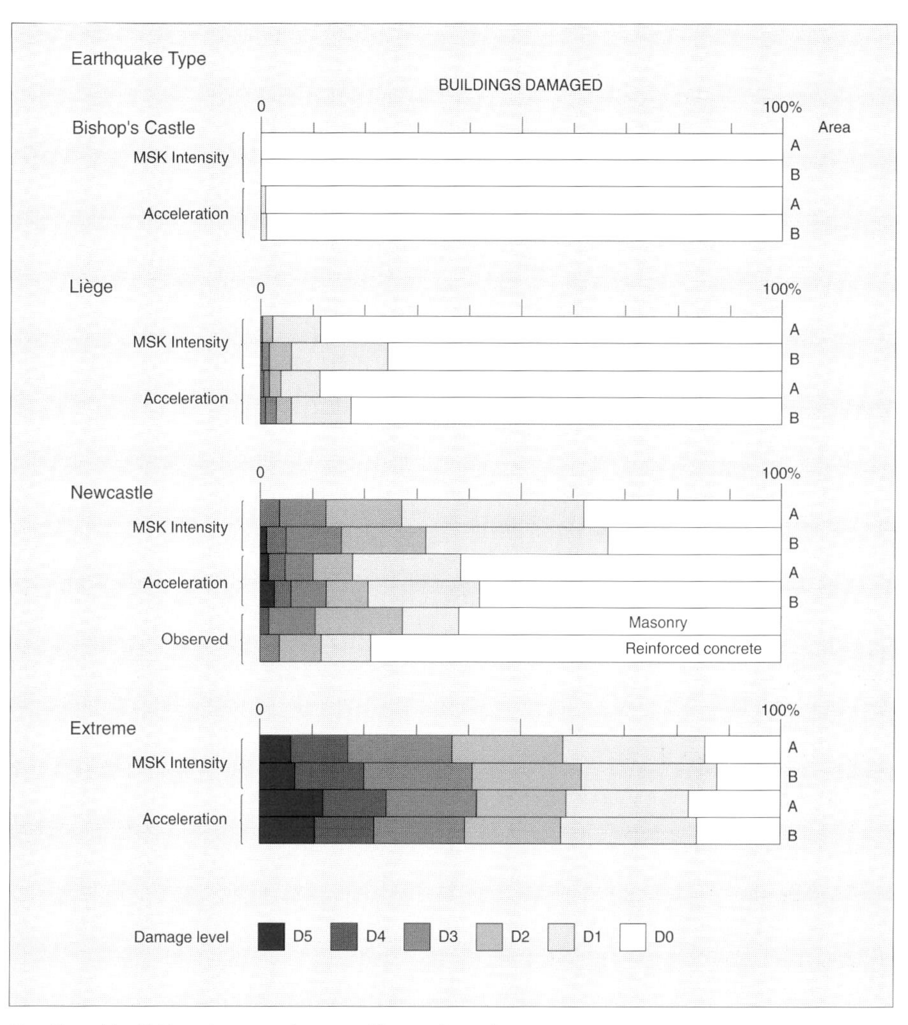

Predicted building damage for specific earthquakes

Earthquake Impact Study Methodolgy

The impact analyses have been carried out by considering a grid overlying the area of interest. In this study a 0.5 x 0.5km grid was used. The geographical information at the centre of each grid was determined so that the number and type of buildings could be estimated.

In order to calculate the cost arising from a specific earthquake the damage in each grid square is calculated using the following steps.

- calculate the value of the seismic hazard parameter using the attenuation relationship and appropriate focal distance;
- calculate the soil amplification if appropriate;
- calculate the ψ value corresponding to the seismic hazard parameter;
- determine the fraction of each building type damaged to damage levels D1 to D5
- calculate the cost of damage.

For all calculations the earthquakes were assumed to occur in the centre of the areas. Two hazard parameters have been used, namely MSK Intensity and peak horizontal acceleration corresponding to a fundamental structural period of 0.2 seconds. Local soil amplifications were applied in the acceleration calculation.

Predicted fatalities

One of the worst consequences of earthquakes causing structural damage is the risk to the population of death and injury. For the purposes of comparing earthquake risk to other risks in terms of loss of life, casualty predictions were made from the estimates of damage from the specific events considered. The method used for human casualty prediction was based on those developed at the University of Cambridge [8].

Between 0 and 500 fatalities with a best estimate of around 10 are predicted for the Liege type earthquake. The actual number of people killed in Liege was 3. Between 100 and 4000 with a best estimate of around 500 are predicted for the Newcastle type event and in fact 12 people were killed. If the earthquake had not occurred on a holiday that number could easily have reached 100. It would have increased again if the housing building stock had been more similar to that of the UK (i.e. 2 to 3 storey masonry). The predicted fatalities, especially the upper estimate, therefore appear to be too high.

Infrastructure Damage

For the four specific earthquakes typical MSK Intensities are V, VI, VII and VIII respectively. For MSK Intensity VI, corresponding to the Liege type earthquake, virtually no damage is expected. For MSK Intensity VIII corresponding to the extreme earthquake, between 5 and 10% cost damage would be experienced by tall chimneys and single span bridges. For the Newcastle type earthquake, MSK Intensity VII is predicted, which indicates about 5% cost damage to masonry chimneys and less than 2% to other facilities. In the Newcastle, Australia earthquake, MSK Intensity VII was generally observed and in that earthquake there was virtually no damage to any non building structures and infrastructure.

Failure of elevated highway in the 1989 San Francisco earthquake

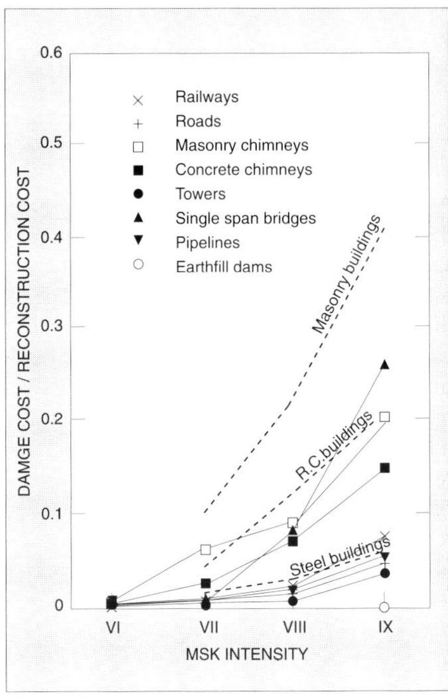

ATC best estimate of damage to non building structures and infrastructure [4]

Predicting fatalities

There are large uncertainties involved in predicting human casualties in earthquakes. Published literature shows that prediction of numbers of fatalities are generally more accurate for more damaging earthquakes (numbers of buildings damaged ≥ D3 of more than 10,000) than for moderate events. This is because at high levels of damage the deaths and injury due to structural damage form a greater percentage of the total than they do at low levels. At low levels of damage to the building stock, as are being considered here, the highest proportion of casualties are caused by damage to elements other than buildings (e.g. garden walls), by non-structural damage to buildings (glazing, cladding, interior contents) and other causes. Predicting human fatalities can therefore only be very approximate.

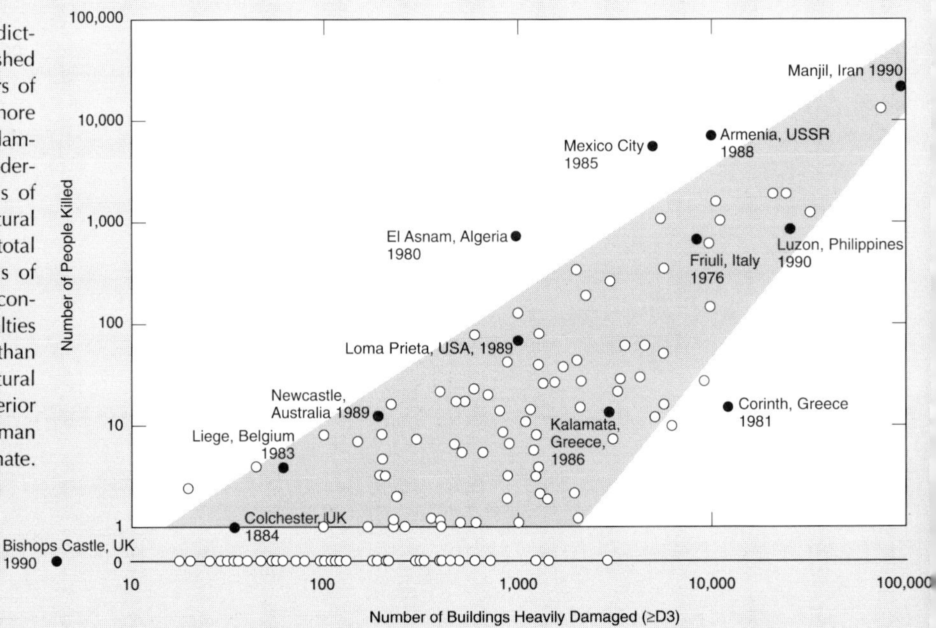

Annual costs from earthquakes

As stated in the introduction risk is defined as he "expected amount of damage due to a specific earthquake or which may occur in any given period of time". It is also stated hat risk is the combination of hazard and vulnerability. This part of the summary presents the calculated risk, in terms of annual cost as a ratio of reconstruction cost, for masonry buildings. The sensitivity of the result to uncertainty of the input parameters has also been examined. In the technical report a similar study is presented for reinforced concrete structures. The relative effects of the uncertainty are similar to that for masonry structures but the annual cost ratios for reinforced concrete structures are calculated to be about a third of those for masonry.

To study the effects of uncertainty in the hazard and vulnerability parameters, a parametric study of the annual cost of masonry buildings subject to the average UK seismicity has been carried out using both MSK Intensity and horizontal acceleration as the hazard parameter. The results of this study are hown here in terms of annual earthquake damage cost as a ratio of reconstruction cost.

MSK Intensity: To study the effect of the uncertainty of the relationship between ψ and MSK Intensity the calculation has been repeated with a variation of ψ to MSK Intensity of $\pm 2\psi$ units (approximately 1 unit of MSK Intensity). As can be seen this variation affects the annual cost ratio by about a factor of 2.5. The variation of the seismic hazard across the country has been calculated by using the locations with the lowest and the highest hazard level in the diagram on page 18. The resulting annual cost ratios are again affected by about a factor of 2.5. This variation is well within the possible band of uncertainty.

Horizontal acceleration: Site soil response effects can be explicitly considered when considering acceleration. As shown here, the resulting annual cost varies from 2 in 100,000 at rock sites up to 16 in 100,000 for a site comprising 5m of stiff clay over bedrock. This variation is significant and has a greater effect than the other uncertainties. Only 5 and 15m deep soil sites have been considered here because they have the greatest effect on low period structures and generally they are more common in the UK than deeper soil sites.

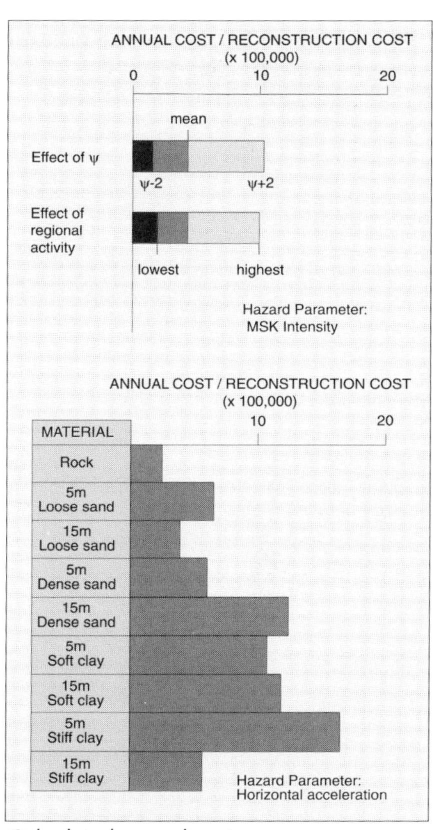

Calculated annual costs

Calculating annual cost

The annual cost calculation is similar to that used in the microzoning analysis for a specific building type. In this case however the calculation is integrated over the full range of hazard values each with its own level of annual rate of occurrence. The diagram below shows the results for damage to masonry structures arising from average seismic hazard in the UK. The calculation is in terms of the peak acceleration response of structure with a fundamental period of 0.2 second for a site underlain by 5m of dense sand. The overall annual cost as a ratio of reconstruction cost is calculated to be 6 in 100,000 and arises mainly from a response acceleration of about 50% of gravity which has an annual rate of occurrence of 5 in 100,000. This implies that a very long period of observation is required before a realistic measure of earthquake damage is obtained.

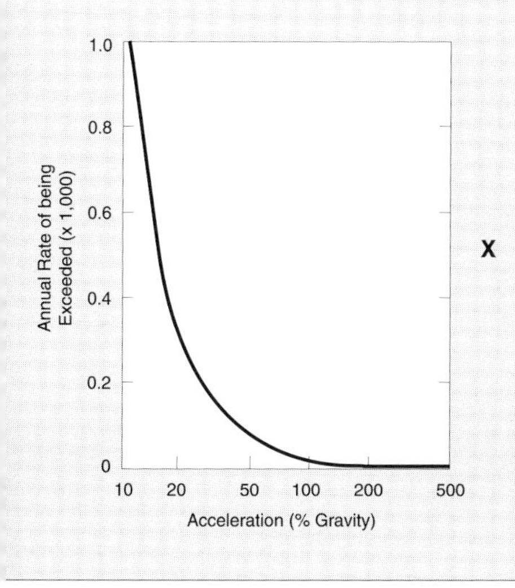

X

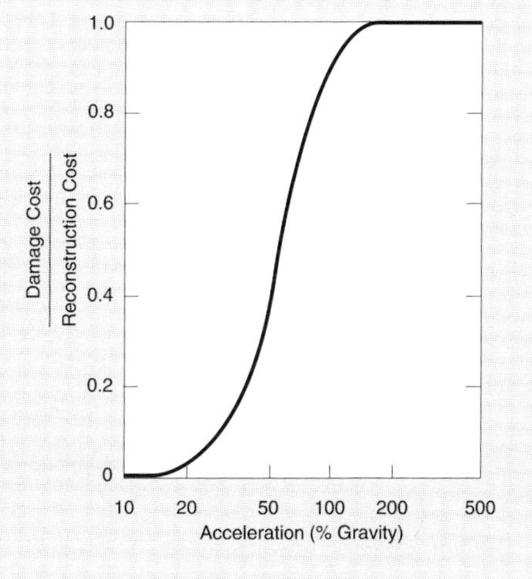

=

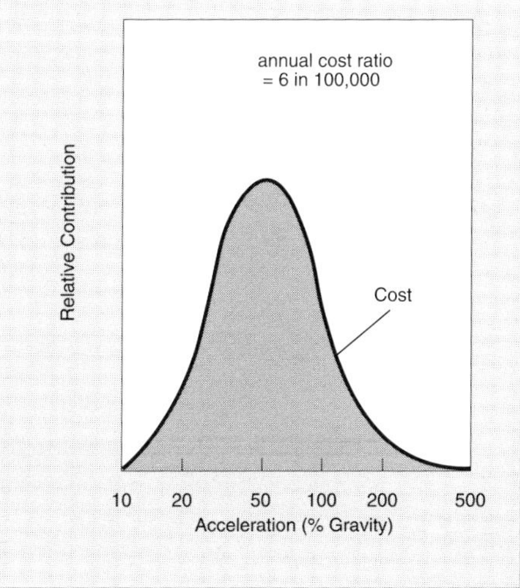

Comparison of earthquake risk to other risks

In order to put the risk from earthquakes in context the risk from other types of natural hazard have been considered.

Financial Cost

The overall annual cost calculations show annual earthquake costs as a ratio of reconstruction cost of around 2 to 5 in 100,000 which corresponds to annual costs in the UK of between £40 million to £100 million. This is significantly higher than that observed in recent years but earthquake losses are very sporadic mainly arising from high losses due to infrequent events.

On the basis of insurance settlements in the past 10 years all types of weather incidents including gales, snow and floods, add up to about £500 million per year (at 1990 prices). In 1990 insurers paid about £500 million in subsidence claims and about £800 million was paid in fire claims.

It follows that the annual estimate of £40 million to £100 million for earthquake losses, while it may well be too high, is still a modest cost compared to observed weather, fire or subsidence claims. It must also be emphasised that these observed figures are not directly comparable with the predicted earthquake cost. If similar prediction exercises were to be carried out for any of these other hazards this conclusion would be reinforced because it is probable that the predicted integrated cost over a long time period would be higher than that observed over the past few years.

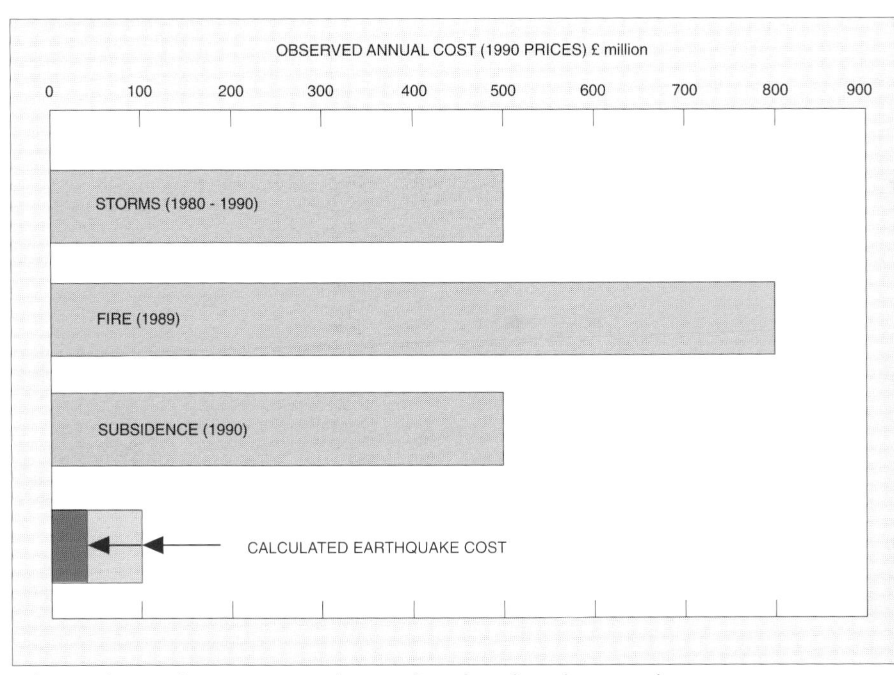

Observed annual cost compared to predicted earthquake annual cost

Risk of Fatalities

In the United Kingdom there are no formal guidelines on the levels of tolerable and intolerable risk to both the individual and to society at large. Statistics of fatalities are maintained however and this diagram shows those observed in the UK from a variety of causes recorded over the 15 to 25 years prior to 1986 [13]. The curves plot the annual rate of an event against the number of fatalities it causes. The results of the predictions of fatalities from the earthquake impact studies are also shown. The predictions for earthquake events are seen be lower than those observed from other hazards but are still quite high for the Newcastle, Australia type event.

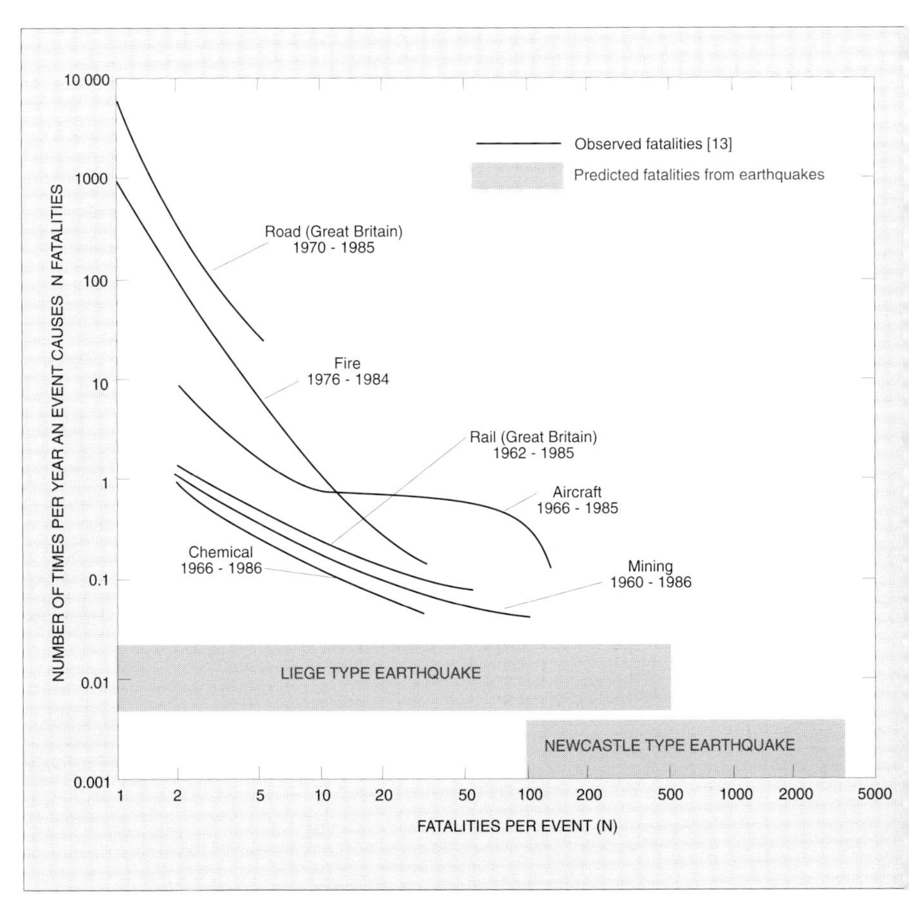

Predicted fatalities in the UK and those observed from other hazards

Conclusions

- The rate of occurrence of earthquakes in the UK is low being less than one hundredth of highly seismic areas such as Japan, Greece and California.

- The variation of seismic hazard across the country is relatively small, being less significant than the effects of the uncertainties of the parameters used in the risk assessment.

- Since there is only a small possibility of an earthquake causing significant damage , the risk to society is sufficiently low not to cause undue concern.

- In the event of a large earthquake (by UK standards) the amount of damage would depend on its location. There is a slight possibility of extensive damage and a limited number of fatalities within an area of a few square kilometres.

- In these circumstances it is not sensible to enact extensive earthquake requirements in the design of conventional structures.

- It would be prudent, however, and incur minimal cost, to subject new buildings to a basic level of earthquake checks. Buildings of national and historical importance should also be checked for robustness to earthquake loading.

- Earthquake loading should be incorporated in the design of structures where the failure of any single element could lead to death or injury of a significant number of people. In addition to nuclear facilities, which are already designed against seismic loading [18], these structures may include some fuel and chemical storage installations, dams and unusual structures such as large sports arenas and chimneys.

Column failure in the 1989 Newcastle, Australia earthquake caused by structural irregularity: Source [27]

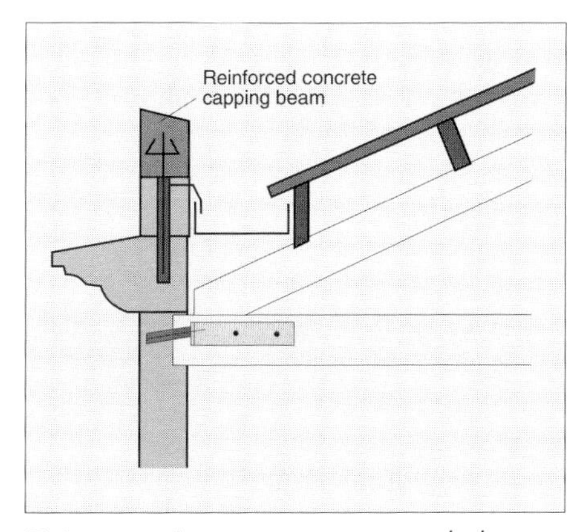

Tying requirements recommended following the 1989 Newcastle, Australia earthquake

Recommendations

- The instrumental monitoring of earthquakes in the UK over the past 20 years has helped our understanding of seismic activity. The detection threshold should be maintained to enable at least magnitude 2 M_L earthquakes to be reliably detected. In addition it is necessary to ensure that large earthquakes, by UK standards, do not exceed the range of the monitoring instruments.

- This Study has shown that some existing structures may have a higher than average vulnerability. As a first step to deciding whether building regulations are necessary, it is recommended that the earthquake risk of a selection of more vulnerable structures be established. These structures would be chosen on the basis of structural irregularity and soil conditions.

- Basic earthquake requirements for conventional structures need to be developed. They should take the form of a checklist for regularity, adequate tying of walls and soil conditions. If any vulnerable features are identified brief design rules, derived from basic good seismic engineering practice, should be given.

- Guidance is required as to how to decide when a structure, or installation, is sufficiently hazardous to require its design to directly consider earthquake loading and, if so, what level of earthquake ground motion is appropriate. The Health and Safety Executive have published many documents concerned with hazardous installations [e.g.. 14] and these should form the basis of this guidance.

References

1. Ambraseys, N.N. 1985. Intensity attenuation and magnitu de intensity relationship for North Western European Earthquakes, *Earthq. Eng. and Structural Dynamics. 13: 733-778.*

2. Ambraseys, N.N. 1988. Engineering Seismology, *1st Mallet -Milne Lecture, Earthq. Eng. and Structural Dynamics. 17: 1-105.*

3. Ambraseys, N.N. 1990. Uniform Re-evaluation of European earthquakes associated with strong-motion records, *Earthq. Eng. and Structural Dynamics. 19: 1-20.*

4. ATC-13. 1985. Earthquake Damage Evaluation Data for California. *Applied Technology council Report ATC-13. Berkeley, California, U.S.A.*

5. Booth, E.D. and Baker, M.J. 1990. Code Provisions for Engineered Building Structures in Areas of Low Seismicity, *Proc. 9th. Eur. Conf. Earthq. Eng. Moscow 1: 269-278.*

6. Browitt, C.W.A., Turbitt, T. and Morgan, S.N. 1985. In vestigation of British earthquakes using the national monitoring network of the British Geological Survey, *Proc. Conf. Earthq. Eng. in Britain. 33-47. Publ. Thomas Telford.*

7. Burton, P.W., Musson, R.M.W., and Nielson, G., 1984. Studies of Historical British Earthquakes, *British Geological Survey, Global Seismology Unit, Report No. 237.*

8. Coburn, A.W., Sakai, S., Spence, R.J.S. and Pomonis, A. 1991. Reducing Human Casualties in Building Collapse, *Martin Centre Report, Cambridge University.*

9. Coppersmith, K.J. and Youngs, R.R. 1986. Capturing uncertainty in probabilistic seismic hazard assessments within intraplate tectonic environments, *Proc. 3rd. U.S. National Conf. Earthq. Eng. 1: 301-312.*

10. Cornell, C.A. 1968. Engineering seismic risk analysis, *Bull. of the Seismological Society of America. 58: 1583-1606.*

11. Dahle, A., Bungum, H. and Kvamme, L.B. 1990. Attenuation modelling inferred from intraplate earthquake record ings, *Earthq. Eng. and Structural Dynamics. 19: 1125-1141.*

12. EEFIT. 1991. The Newcastle, Australia Earthquake, *Publ. Earthquake Engineering Field Investigation Team. Inst. Structural Engineers., London.*

13. Fernandes-Russell, D. 1988. Societal risk estimates from historical data for UK and worldwide events, *Research Report 3. Environmental Risk Assessment Unit, Univ. of East Anglia, Norwich.*

14. Health and Safety Executive 1989. Risk Criteria for land-use planning in the vicinity of major industrial hazards. *Health and Safety Executive, Publ. HMSO, London.*

15. Heidebrecht, A.C., Henderson, P., Naumoski, N. and Pappir J.W. 1990. Seismic response and design for structures located on soft clay sites, *Canadian Geotechnical Journal 27: 330-341.*

16. Hwang, H.H.M. and Jing-Wen Jaw. 1990. Probabilistic Damage Analysis of Structures, *ASCE, Jour. Structural Eng. 116: 1992-2007.*

17. Jongmans, D. and Campillo, M. 1990. The 1983 Liege Earthquake: Damage Distribution and Site Effects, *Earthquak Spectra 6: 713-737.*

18. Mallard, D.J., Higginbottom, I.E., Muir Wood, R. and Skipp B.O. 1991. Recent developments in the methodology of seismic hazard assessment. *Proc. ICE Conf. Civil Eng. in th Nuclear Industry. UK.*

19. Melville, C.P.M. 1985. The geography and intensity of earth quakes in Britain -the 18th. Century, *Proc. Conf. Earthq. Eng in Britain: 7-24. Publ. Thomas Telford.*

20. Muir Wood, R. 1989. Fifty million years of 'Passive Margir deformation in North West Europe, *7-36 in S. Gregersen an P.W. Basham eds.,* Earthquakes at North Atlantic Passive Margins: Neotectonics and Postglacial Rebound". *Kluwer academic Publishers, Dordrecht.*

21. Musson, R.M.W. 1990. Fatalities in British Earthquakes, *British Geological Survey, Global Seismology Research Grou WL/90/26.*

22. NEHRP.1988. National Earthquake Hazard Reduction Programs. Recommended provisions for the development o seismic regulations for new buildings, *Building Seismic Safety Council. Washington DC.*

23. Principia Mechanica Limited1982 British Earthquakes, *Report prepared for Central Electrical Generating Board, British Nuclear Fuels Ltd. and South of Scotland Electricity Board.*

24. Seed, H.B. 1986. Influence of Local Soil Conditions on Ground Motions and Building Damage during Earthquakes *8th. Nabor Carrillo Lecture. Mexican Society for Soil Mechanics, Mazatlan, Mexico.*

25. Soil Mechanics Ltd: 1982. Reassessment of UK Seismicity, *Report prepared for Central Electrical Generating Board, No. 7984.*

26. Spence, R.J.S., Coburn, A.W., Sakai, S. and Pomonis, A. 199 A parameterless scale of seismic intensity for use in seismi risk analysis and vulnerability assessment, *Proc. SECED, Con Earthquake Blast and Impact, 19-30. Publ. Elsevier*

27. Working magazine 1990. *New South Wales Public Works Department, Sydney, 4: No. 8, 9-12*

Acknowledgements

All the members of the project team wish to acknowledge the Department of the Environment particularly Brian Marker, and all members of the Steering Committee for their continued encouragement and helpful criticism throughout the project. Particular thanks must be made to Robin Adams, Professor Ambraseys, David Mallard, Roger Musson, and Brian Skipp for their assistance.

Members of the Steering committee who guided the Study were from the following organisations:

Department of the Environment, London
Her Majesty's Inspectorate of Pollution, London
Building Research Establishment, Hertfordshire
Ministry of Defence, Reading
International Seismological Centre, Berkshire
British Coal Corporation, Staffordshire
Scottish Hydro-Electric plc, Perth
British Gas Corporation, London
Health & Safety Executive, Merseyside
Scottish Power plc, Glasgow
Nuclear Installations Inspectorate, HSE, Merseyside
Scottish Development Department, Edinburgh
AEA Technology, Warrington
Imperial College, London
British Geological Survey, Edinburgh
Welsh Office, Cardiff
BP International Ltd, London
Nuclear Electric plc, Gloucester
Soil Mechanics Associates, Berkshire

This summary report was prepared by:
Ove Arup and Partners
13 Fitzroy Street
London W1P 6BQ
United Kingdom
Tel: 071-636 1531
Fax: 071-465 2121

Printed in the United Kingdom for HMSO
D296023 C15 6/93

Technical Report

The Technical Report describing this Study consists of about 100 pages each of text, figures and tables. It details and justifies all of the assumptions used in the Study, shows how the results have been calculated and includes a Fortran listing of the seismic hazard computer program.

In addition to presenting the earthquake catalogues and the derivation of seismic hazard at various locations around the country the Technical Report includes a detailed sensitivity study of the hazard calculations. The effects of local soil conditions, mining induced seismicity and the likelihood of liquefaction are also discussed. The vulnerability of various types of structures and the earthquake impact studies, referred to as microzoning studies in the Technical Report, are described in detail. Structures with high consequence of failure are also discussed and preliminary rules for the levels of earthquake ground motion for design are presented.

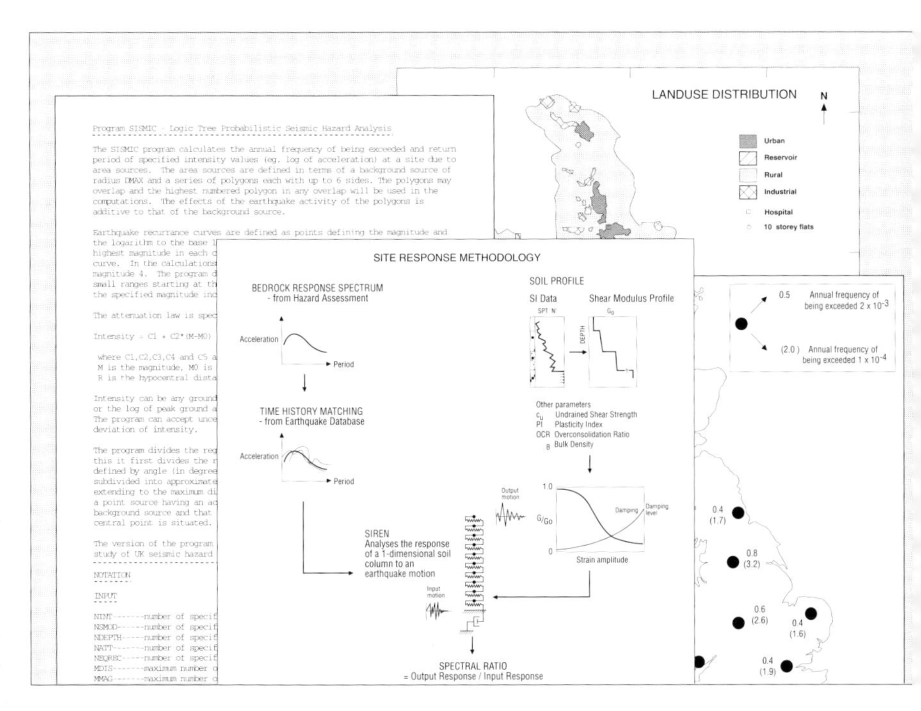

Oasys UKCAT, UK earthquake catalogue

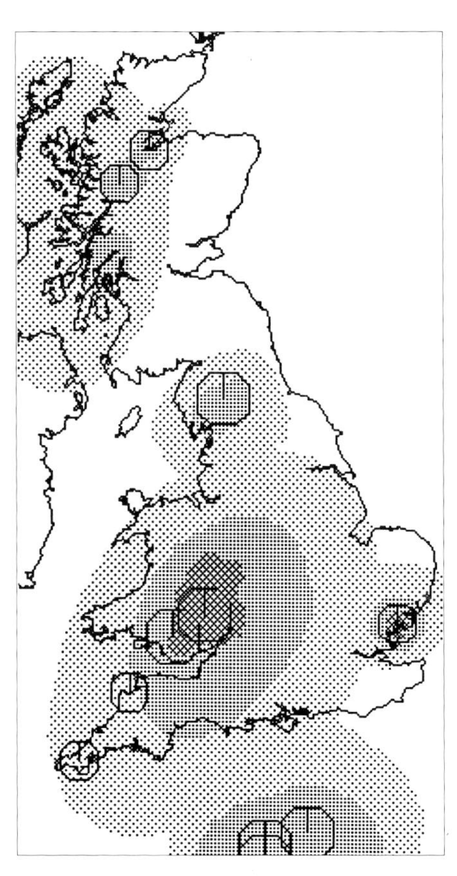

This computer program contains a catalogue of instrumentally recorded earthquakes $M_L >$ 2 since 1979 and macroseismically recorded earthquakes since 1000. The data include isoseismal maps since about 1700.

The program enables listings of the catalogues and plots epicentres and isoseismal maps. It also displays maps of intensity recurrence since 1800 and for the 19th. and 20th. centuries. It will run on most IBM compatible personal computers

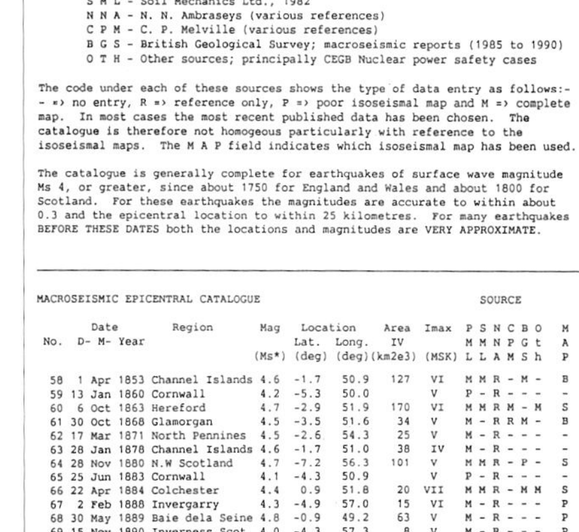

MACROSEISMIC EARTHQUAKE CATALOGUE FOR THE UK

This catalogue has been assembled from the following sources

P M L - Principia Mechanica Ltd., 1982
S M L - Soil Mechanics Ltd., 1982
N N A - N. N. Ambraseys (various references)
C P M - C. P. Melville (various references)
B G S - British Geological Survey; macroseismic reports (1985 to 1990)
O T H - Other sources; principally CEGB Nuclear power safety cases

The code under each of these sources shows the type of data entry as follows:- - => no entry, R => reference only, P => poor isoseismal map and M => complete map. In most cases the most recent published data has been chosen. The catalogue is therefore not homogeous particularly with reference to the isoseismal maps. The M A P field indicates which isoseismal map has been used.

The catalogue is generally complete for earthquakes of surface wave magnitude Ms 4, or greater, since about 1750 for England and Wales and about 1800 for Scotland. For these earthquakes the magnitudes are accurate to within about 0.3 and the epicentral location to within 25 kilometres. For many earthquakes BEFORE THESE DATES both the locations and magnitudes are VERY APPROXIMATE.

MACROSEISMIC EPICENTRAL CATALOGUE

No.	Date D- M- Year	Region	Mag (Ms*)	Location Lat. (deg)	Long. (deg)	Area IV (km2e3)	Imax (MSK)	SOURCE PML	SML	NNA	CPM	BGS	Oth	MAP
58	1 Apr 1853	Channel Islands	4.6	-1.7	50.9	127	VI	M	M	R	-	M	-	B
59	13 Jan 1860	Cornwall	4.2	-5.3	50.0		V	P	-	R	-	-	-	-
60	6 Oct 1863	Hereford	4.7	-2.9	51.9	170	VI	M	M	R	M	-	M	S
61	30 Oct 1868	Glamorgan	4.5	-3.5	51.6	34	V	M	-	R	R	M	-	B
62	17 Mar 1871	North Pennines	4.5	-2.6	54.3	25	V	M	-	R	-	-	-	-
63	28 Jan 1878	Channel Islands	4.6	-1.7	51.0	38	IV	M	-	R	-	-	-	-
64	28 Nov 1880	N.W Scotland	4.7	-7.2	56.3	101	V	M	M	R	-	P	-	S
65	25 Jun 1883	Cornwall	4.1	-4.3	50.9		V	P	-	R	-	-	-	-
66	22 Apr 1884	Colchester	4.4	0.9	51.8	20	VII	M	M	R	-	M	M	S
67	2 Feb 1888	Invergarry	4.3	-4.9	57.0	15	VI	M	-	R	-	-	-	P
68	30 May 1889	Baie dela Seine	4.8	-0.9	49.2	63	V	M	-	R	-	-	-	P
69	15 Nov 1890	Inverness,Scot.	4.0	-4.3	57.3	8	V	M	-	R	-	-	-	P

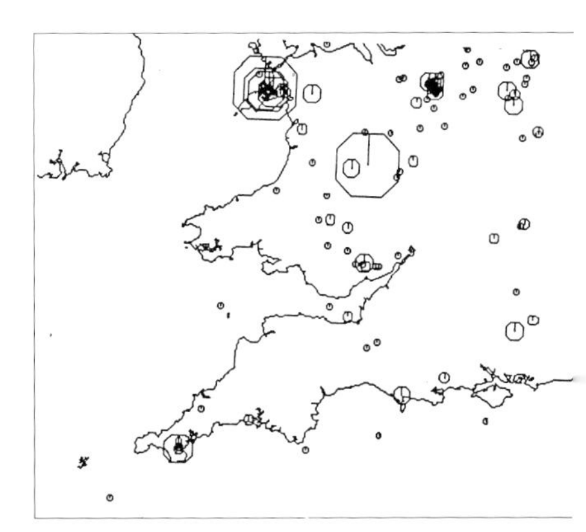

Availability

The Technical Report costs £250 and available from:

Dr. J.W. Pappin
Ove Arup and Partners,
13 Fitzroy St.,
London W1P 6BQ,
United Kingdom.

For an additional cost of £150 a copy o *Oasys* UKCAT is included.